AF618538

Project Management at the Edge of Chaos

Alfred Oswald • Jens Köhler • Roland Schmitt

Project Management at the Edge of Chaos

Social Techniques for Complex Systems

Alfred Oswald
IFST – Institute for Social Technologies GmbH
Stolberg, Germany

Jens Köhler
BASF SE
Ludwigshafen, Germany

Roland Schmitt
Weinheim, Germany

This German-English translation has been carried out with the help of Yannick Oswald and proof-read by Jeannie Graham. Title of the orginal German Edition: Projektmanagement am Rande des Chaos © Springer-Verlag Berlin Heidelberg 2016

ISBN 978-3-662-48260-5 ISBN 978-3-662-48261-2 (eBook)
https://doi.org/10.1007/978-3-662-48261-2

Library of Congress Control Number: 2017964707

Printed on acid-free paper

This Springer imprint is published by the registered company Springer-Verlag GmbH, DE part of Springer Nature
The registered company address is: Heidelberger Platz 3, 14197 Berlin, Germany

Foreword by Heinz Schelle

Some time ago, when Alfred Oswald asked me to write a foreword for this book, I agreed without hesitation. I was very familiar with the book "The Collective Mind Method. Project Success through Soft Skills" (Note: English translation of the German book title), which he had published with Jens Köhler in 2009. I had discussed it at length in the journal "projektMANAGEMENT aktuell" (issue 5-2011) and still consider it one of the most important German project management books of the last few years with its central message being that people should be used in projects according to their strengths. To accomplish this task, a comprehensive set of instruments is offered.

The current aim of the authors, now including Roland Schmitt, is also to provide appropriate social techniques. This time it is the authors' intention, to master the complexity of projects embedded in social interactions. For a long time in our discipline, it was almost exclusively about getting a grip on the complexity of developing technical systems. The most impressive example is probably the contributions that NASA and the United States Department of Defense have provided for project management development. In this context we can consider network planning, and in the context of the planning and development of programs, such as the Apollo program, sophisticated configuration management, which in some cases is even considered an engineering discipline. "New Taylorism" as a term for this import from the United States was by no means entirely unjustified. In any case, man as acting subject in projects, is not present in these concepts.

This very biased, even rightly called, technocratic approach, has long dominated our view of projects, perhaps for too long. Even today, it still dominates in many textbooks. It was relatively late that the perception expanded to projects as social systems. One of the protagonist in German-speaking countries was the Austrian Consulting Group Neuwaldegg. In the meantime, our knowledge of social systems has expanded considerably. Organizational psychology, behavioral economics, Synergetics and neuroscience can be mentioned here, as a scientific disciplines that have greatly contributed to this new perspective.

The undertaking of the three authors – to provide the tools for "optimum tuning of social interactions" and the management of complex systems, based on a viable theory and model with their motto: "nothing is more practical than a good theory", thus has considerable potential. The authors succeed in making complex material easily understandable, interspersing with a variety of stories and extensive application examples.

The authors regretfully note that the subjects of complexity and self-organization in social systems are not covered at school or university or during vocational training. And most certainly not in politics. Current discussions on how to deal with the refugee problem clearly demonstrates this. Thinking in simple linear cause-effect relationships predominates. Every day, we can see that measures decided and implemented by governments ad hoc in blind activism, do not even consider even slightly predictable and relatively short-term side effects.

It is therefore imperative that the "method gap in dealing with complexity" is closed. This is the aim of the authors with this work. I congratulate the team and wish the book and the ideas contained within it, a wide distribution.

Univ.-Prof. Dr. Heinz Schelle
Honorary Chairman of GPM German Association for Project Management e.V.

Authors' Preface to the English Edition

This book is a translation of the German edition of our book "Projektmanagement am Rande des Chaos", published by Springer-Vieweg in 2016 and 2017, respectively. Despite the vast array of existing management literature, we nevertheless see a gap in the area of management dealing with theories and models in complex contexts, e.g. as required for complex projects. We have therefore taken a completely new approach and transferred the fundamental principles of the universal phenomenon of self-organization as described in the theory of Synergetics, from the natural sciences to leadership of organizations and projects. We believe that this approach will provide deep, new insights to the reader, with only a few parameters, the so called setting, control and order parameters, very well suited to determine the coarse-grained behavior of social systems such as organizations and projects. This will surely help move the focus away from usual nitty-gritty details, to a few, but important quantities, which will bring the reader directly to a solution for everyday problems at work in industrial and public organizations. Concentrating on micro-tier details and applying straight-forward linear concepts (as with Taylorism) is no longer sufficient to solve complex real world problems. Linearizing methods are simply outdated for the management of complex environments! We are sure that you, the reader, perceive this every day!

Individuals and their interactions are the center of our attention. We make transparent their behavior, motives, needs, temperament and culture, including their mutual dependencies, and we state how an extremely practical framework can be built upon an integral network of theories and models. "Nothing is more practical than a good theory" is what we genuinely believe! We hope that the reader can apply the material we have presented immediately in their field of business. If any reader feels that our management book is more abstract than an English management textbook, we encourage them to say out loud: "Nothing is more practical than a good theory" and then just apply it – to see an immediate benefit.

The book also contains entertaining dialogues and stories to elucidate our message. We have adapted some of the personal names and cultural specifics in the German edition to suit the English-speaking world.

With regard to the cited literature, we have used both English and German sources. We apologize if some of the sources are only available in German.

We truly believe that this book will enrich the English-speaking market of management literature and hope the reader enjoys gaining insight and becoming familiar with the material presented in the book.

Stolberg, Mutterstadt and Weinheim an der Bergstraße
November 2017

Alfred Oswald
Jens Köhler
Roland Schmitt

Authors' Preface to the German Edition

Not another book on complexity and this time with a focus on project management! And to top it all, it deals extensively with models and theories! However, it is a book that is necessary and even imperative: In the current and future project world, we are dealing with interconnected, feedback systems, meaning we have arrived directly in the realm of complexity: It creates self-organization and thus completely new structures, intended or unintended, stable or chaotic. Complexity is not a phenomenon of our time, but is part of nature and furthermore, the basis of any form of life.

Unfortunately, the topics of complexity and self-organization in project management are far from having been covered widely, whether at school, vocational training or university. However, practice requires that we deal with complexity now: Think for example, of a composite of interconnected companies or a multi-component, networked IT system: Changes occurring in one place, suddenly result in unexpected changes elsewhere. A linear cause-effect analysis is no longer valid. Now, to a large extent, value creation in industry, occurs in projects. High values can be created, as well as rapidly destroyed. This book will contribute to closing the method gap for dealing with complexity, and further contribute to minimizing the value-destroying component in the everyday life of projects and help to lead the value-creating component to new, outstanding project solutions.

Complexity lies not only in the technologies used, but also in social interactions. Regulating and controlling these interactions by means of appropriate social techniques is the subject of this book and demands an engagement with theories and models. How do we wish to impart this information? One of the key messages of this book is: There is nothing more practical than a good theory.

Let us directly adopt this sentence and use a simple theory, the metaphor of the "iceberg", as an elucidation of the very basis of the book: An 1/8 of an iceberg is visible above the water, but 7/8 is invisible below the water surface. Although this 7/8 is indeed invisible, it is by no means ineffective, with potentially fatal consequences for those at sea. When this metaphor is applied to the social system "project", "traditional" project management methods and processes, which are above the "iceberg", represent smaller, visible events – but the larger and more dominant part remains below the "waterline", and this represents social factors.

The true purpose of this book is to make projects successful by the optimum adjustment of social interactions in a work world, which is increasingly dominated by knowledge work. This can be achieved by a framework of networked theories and models based on the latest scientific findings and adapted for project practice. For this purpose, among others, we use the latest research results in neuroscience and Synergetics, which are both topics of complexity research.

After a general introduction in Chap. 1, in Chap. 2 we address the relationship between social techniques and complexity. Fundamental definitions, such as "What is complexity?", will be given. In Chap. 3, we discuss the foundations for complexity regulation options. With the fundamentals and possibilities of complexity regulation in Chap. 4, options for leadership in complex social systems are outlined, derived from models and based on concrete examples. With the help of the framework developed in the previous chapters, in Chap. 5, we compare known management systems and check their basic assumptions. Finally, Chap. 6 draws a conclusion and an outlook: This may encourage you to use the contents of the book and thereby sharpen your skills in dealing with complexity.

No less important than the main text in Chaps. 1 to 6, are the annexes: To achieve stringency and better follow the guiding thread for understanding necessary descriptions of important models and theories. Depending on personal knowledge background, it may also be useful to turn to look at the annexes first.

Examples in the book should also not be forgotten: For better understanding and to support the transfer of knowledge into practice, the following examples have been included:

In Chaps. 2 and 3, among other examples, you will be accompanied by Tobias Ehrlich and Heiner Priesberg, both employees of the fictional company MedicalFit, who explain certain content in depth, based on partly exaggerated dialogues.

In the key Chap. 4 "Leadership in Complex Social Systems" these dialogues flow into more extensive, accompanying application examples.

In Chap. 5 "Consequences for Management Systems", three detailed, practice-relevant examples of fictitious companies are used for deepening knowledge.

For further illustration purposes, in several passages, various working examples taken from workshops have been incorporated into the text.

For better traceability, the text is structured in such a way that many cross-references and content clues are included. In addition, we have deliberately incorporated redundant information, so that content can be branded in the memory more easily.

Chapters 1 to 4 are preceded by quotations from the works of William Shakespeare and the crime author Fred Vargas. These aim of these quotes is to set the scene for the following text and invite associative play of thoughts.

We hope with the book and the associated different ways of approaching complexity, to provide a tool in your hands, which increases your knowledge and practice, based on your experience so far and helps you brace yourself for future situations in a complex environment.

We hope that this book will also contribute to diffusing complexity and self-organization on a wider scale.

Stolberg, Mutterstadt and Weinheim an der Bergstraße

Alfred Oswald
Jens Köhler
Roland Schmitt

Acknowledgements

A book such as this is always preceded by continuous and intensive professional discourse in a stimulating practice environment. Therefore, we would firstly like to thank Prof. Dr. Heinz Schelle, a doyen of project management in the German-speaking community, for his continued interest in our work, and in particular for the foreword to this book. We would like to thank him and the GPM e.V. (German Association for Project Management) for the professional cooperation and enabling our ideas and concepts to be published.

We are also genuinely grateful to Ms. Susanne Schwarzer for her knowledgeable contributions to the integration of theory and practice.

We would like to take the opportunity to sincerely thank Dr. Thomas Lorenz, BASF SE for the many years of support and numerous valuable discussions on project and knowledge management.

We would like to express our thanks to Dr. Hannelore Weber, Björn Decker, Dr. Gerald Lippert and Jan Meyer (all BASF SE) and Dr. Christian Lennartz (trinamiX GmbH), for many years of exciting discussions on topics in the field of project world.

In addition, we would like to thank Hermann Engesser, Dorothea Glaunsinger and Gabi Fischer, Springer Verlag, Heidelberg, very much, for the many years of professional collaboration.

And last but not least, we would like to thank our families. For several years, they have been the stimulating force and support, to enable us to be able to actively push forward our objective, the development of project management. The insights we have gained are also relevant and interesting for the private, as well as the business environment.

Contents

About the authors

Alfred Oswald IFST – Institute for Social Technologies GmbH, Stolberg, Deutschland
earned his doctorate in Theoretical Physics at the RWTH Aachen University. He is Managing Director of the IFST-Institute for Social Technologies GmbH consulting institute, responsible for project management, transformation and innovation management. He is head of the special interest group for Agile Management of the GPM (German Association for Project Management e.V.). His field of work is the efficiency and effectiveness of organizations through innovative social technologies. He has many years of experience in the management of innovative and complex projects, as well as in the transformation of project-oriented organizations into high-performance organizations.

Jens Köhler BASF SE, Ludwigshafen, Deutschland
received his Diploma in Physics from Bonn University and already began tackling the subject of complexity in his doctorate. He focuses on the digitalization in research and development. His specialty is the regulation of social complexity to lever the efficiency and effectiveness of project teams, being inevitable for successfully forming the digital transformation.

Roland Schmitt Weinheim (an der Bergstraße), Deutschland
first gained experience in managing business and software development processes after his studies in Electrical Engineering. For more than ten years he has managed IT projects in the logistic and transport sector at DB Systel GmbH in Frankfurt am Main. His postgraduate management studies at the University of Portsmouth in the UK focused his attention on the psycho-social aspects of cooperation between people. Since then, his research interest has been the transfer of theoretical findings of self-organization and neuroleadership into professional management practice.

1 Introduction and Motivation

...Things growing are not ripe until their season:
So I, being young, till now ripe not to reason;...

William Shakespeare, A Midsummer Night's Dream

We cannot close our minds to the professional handling of complexity: Projects are increasingly becoming embedded within an incomprehensible and unpredictable context. Context perceived this way, is usually associated with complexity. With this in mind, the mastery of complexity is a necessary precondition for successful project management in the twenty-first century. Additionally, we can assume that the boundaries between project management and general management are increasingly disappearing and that the tools presented in this book are applicable to all kinds of management.

Admittedly, the title "Project Management at the Edge of Chaos" is supposed to grab attention, but at the same time, it represents the contentual design of this book.

During our search for management tools, our aim was to avoid being misguided by stereotypical assertions like "Complexity is different to complicatedness", "Complexity only exists in nature", "Complexity is bad, simplicity is the way", "Get out of the complexity trap", "Complexity must be reduced" or "Complexity is subjective", to name just a few.

Instead our aim was to apply insights from complexity science, physics, neuroscience, psychology and the social sciences to elaborate the field of management. In doing so, we wanted to simplify these insights, but only as long as the insights were not distorted. Likewise, we endeavored to create an interdisciplinary network of theories and models, rather than keep different disciplines as separate individual theories and models. For instance, it is necessary to interpret the insights from neuroscience and the natural-scientific theory of self-organization in light of practice-oriented cases of management,

A. Oswald et al., *Project Management at the Edge of Chaos*,
https://doi.org/10.1007/978-3-662-48261-2_1

and additionally to demonstrate which neuroscientific insights are of importance in the self-organization of social systems, and vice versa.

We emphasize the integral networking of different theories and models, especially because this results in completely new insights and is related to a higher level of quality of management design.

In complex projects, this interdisciplinary network of theories and models generally replaces operations based on Best Practices and standardized methods. However, theories and models are only effective in practice if they are suitable for successfully shaping the world. We will show that the deliberate application of theories and models in project management enables the acquisition of comprehensible experience. Experience acquired this way is the key to developing intuition: The application of theories and models develops intuition in a comprehensible and verifiable way. This is a new approach which reveals intuition to be a professional way of handling complexity. In complex and chaotic systems, such designed and trained intuition is probably the only way to act professionally.

The formation of personal experience based on this integral network of theories and models, has, in combination with professional intuition obtained, the effect of fostering holistic and systemic thinking. It is another crucial precondition in order to be able to take action within complex contexts and systems. To the best of our knowledge, holistic and systemic thinking are present if:

- an intuitive feeling manifests for the networked interaction of system elements and respective emergent system structures,
- a "Big Picture" of the system intuitively appears in one's mind,
- the possibilities and limits of one's own abilities in complex systems are recognized attentively and mindfully.

Therefore, within this book, we address the following key questions:

- What is complexity and what creates complexity? Which systems are able to show complex behavior? How do we recognize complexity?
- When do value-creating and value-destroying complexities occur? How do networked (social) interactions occur?
- Which possibilities exist for regulating and absorbing complexity? Do instruments for regulating and absorbing complexity have to be complex, too?
- Which possibilities and limits exist for the leadership of complex systems? Which role does self-organization have in this context? Which role does intuition have while regulating and absorbing complexity?
- Which theories and models are applied to regulate complex systems? What does the resulting network of theories and models for managing complex projects look like?

2 Social Techniques and Complexity

It's a pity thoughts don't have names, isn't it? You could call them up, and they'd come and lie down at your feet, crawling on their bellies.

Fred Vargas, A Climate of Fear, English by Siân Reynolds

In this chapter we forge a bridge between our fundamental assumptions and beliefs, the "Big Picture" which underlies our project management approach, and our understanding of complexity and the related consequences.

We begin with a discussion of the terms "theory and practice" and "social technology and social technique" and outline, from the perspective of project management, the essential questions, which we intend to answer in this book

Within the section, "Attentiveness, Mindfulness and Wisdom", we show that the approach we propose contributes to an integral perspective of the world.

The "Big Picture" describes an operational framework, which enables the repeated synchronization of theory and practice, to be able to find, inside complex projects or environment, the way to the goal.

In the section "En Route to Complexity", we make an important distinction between drivers of complexity and domains of complexity. In addition, we introduce value-creating complexity and value-destroying complexity.

Subsequently, we link complexity to "blind mental spots" like uncertainty, suspense and risk.

The final section "Philosophy of Complexity" summarizes the implications of a "world shaped by complexity" and lays the foundation for all further chapters.

A. Oswald et al., *Project Management at the Edge of Chaos*,
https://doi.org/10.1007/978-3-662-48261-2_2

2.1 From Negative Words and Basic Assumptions

That's you, is it, Queen Mathilde, making cruel jokes?
Yes, it's me, Charles. I'm getting my retaliation in first. You know what they say: 'If you want peace, prepare for war.'

Fred Vargas, The Chalk Circle Man,
English by Siân Reynolds

Even though terms like "complexity" or "chaos" are on everyone's mind, sometimes it's better to avoid the term complexity in conversations. The higher the represented hierarchy, the less welcome this word is. Complexity, not only indicates unpredictability, but also "we do not want this", "complexity is bad"; in a nutshell, complexity is something to avoid: Complexity emerges where bad management exists.

We will demonstrate that complexity is not always "bad", nor is it due to "poor management", but good management distinguishes itself by utilizing the value-creating potentials of complexity for its goals: Since complexity is the foundation of our being and therefore of any evolution.

Coming close second to this aversion, is the word "theory". Phrases like "we do not want to theorize" or "the whole thing should be practical" are bandied about. If the worst comes to the worst, "best practices" are recalled and people want to rely on them.

There is Nothing more Practical than a Good Theory

Priesberg encounters his colleague Ehrlich in the hallway after a fruitless team meeting. Priesberg seems tired and approaches him: "The whole day long, I have tried to convince my project leader of the fact that his best practices should be supported by theory, especially because they have been transferred from another project and in my view are not useable in the current project. He increasingly treats me like a theorist in an ivory tower– both myself and my effectiveness are undermined, despite the fact that I am one of the backbones of the team."…"and there is nothing more practical than a good theory," Ehrlich cups his hands around his mouth to magnify his voice, and shouts after the project leader as he rushes past them without a word of greeting.

We claim that both theories and individual models require practical value. They need to be useful in practice and they need to be testable in practice. We are guided here by the basic assumption that "There is nothing more practical than a good theory". Theories are nothing more or less than deliberately structured thoughts. And thoughts contain images, therefore models, of the perceived practice. Consciously perceived models are the basis for profound thoughts and subsequently profound theories. Although we value a constructivist approach [1] to the wide range of individual perspectives on the world, it does not lead us to assume that all thoughts are equally appropriate for representing reality. Therefore, we do not believe that all theories and models are equally "good". Different theories and models of the same practice usually differ significantly in terms of their practical relevance. Theories and models are only of practical relevance if they are adapted to shape the world successfully.

We do not want to understand theory only as form of discourse. Nevertheless, we regard it of importance that theories and models are selected via discourse and the best possible theory, in respect of a practical issue, is further developed. It should be possible to derive guidelines for future activities from a good theory and models.

Use Theories and Do Not Ignore them!

Ehrlich approaches his colleague Priesberg: "Nowadays, anyone who fails to come to grips with the value of models and theories, is squandering their future. A few years ago, when project environments were static and the few theories that there were, had been developed as best practices, like the preparation of a network plan; deeper reflections were not necessary. Nowadays, this is not possible, because project environments change abruptly, meaning that best practices often lose their validity when introduced into new contexts. In order to sufficiently understand and govern the environment, tools are required. With the help of these tools, your project leader is able to recognize how his uncooperative behavior is potentially damaging to cooperation within the team and promotes value-destroying complexity: If your role is undermined, professional expertise will not be fully embraced and crucial aspects of the solution to the problem will go unheard. Now, the whole, is less than the sum of its parts."

That does not mean that we have an unconditional belief in the benefits of evidence-based management. We believe that both qualitative and quantitative methods are valid for verifying a theory or a model. We also believe that quantitative methods (e.g. statistics) are no guarantee for meaningfulness, but we are led by the basic assumption that evidence, solely based on statistical methods, is not of value without a practical theory.

Furthermore, we believe that the separation of social and natural sciences may have many practical implications, but should not lead to behavior where common characteristics are not actively searched for. For this reason, the content of this book is grounded on the following fundamental axiom: "Complexity is a universal phenomenon, the basic principles of which, can be applied to all objects or forms of existence". Out of this, our working principle emerges, applying natural scientific principles of complexity to the social discipline of project management.

The sub-title of this book is "Social techniques for complex systems". Here we draw on Beinhocker's understanding of the term, which he outlined in his excellent book "The origin of wealth" [2]:

"Social Technologies (STs) are methods and designs for organizing people in pursuit of a goal or goals." Social technologies are understood as an addition to physical technologies. Physical technologies denote the classical field of technologies in the natural and engineering sciences. Figure 2.1 elucidates this perspective with the help of the "iceberg" model: 7/8 of the factors determining our lives can be found beneath the water surface and only 1/8 of these factors lie above the water surface. Social technologies deal with the explanation of "social factors" and their interplay with "physical factors". By social techniques, we mean concrete forms of social technologies. Hence, we understand project management as a social technology comprised of several social techniques, thus models and methods for leading, planning, team-building etc.

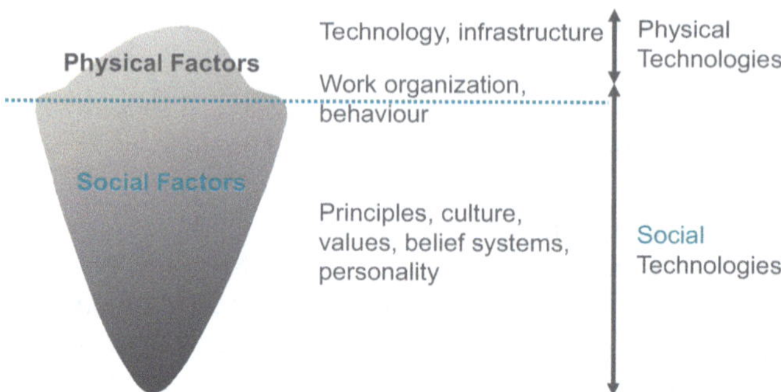

Fig. 2.1 Social Technologies in the iceberg model

The term social technique was recently revisited by the organizational psychologist Lutz von Rosenstiel in the "Mit Sozialtechniken 'Prozessverlusten' entgegenwirken" interview [1] [3] and also used in the book by Lars Vollmer [4], which is well worth reading. This is all the more surprising since the terms "social" and "technique" are perceived as contradictory by some. The sociologist Habermas once downgraded the work of his colleague Luhman by announcing that Luhmann was pursuing social techniques for his system theory, by adopting natural scientific ideas. The term "social engineering" previously created by the philosopher Popper, suffered a similar fate and unfortunately is nowadays often used in the context of manipulative activities.

We want to understand these concepts in their original intention: Social techniques serve to master complex systems, or as we will discuss later on, serve to regulate them. We will examine our understanding of the terms complexity and complex in more detail in the next section. By using a system, we will come to understand an aggregate of elements which relate to and interact with each other in such a way that they are perceived as a unity and, in this respect, delimit themselves against the surrounding environment, their context. Examples of systems are: social systems like teams, organizations and societies, natural systems like humans, animals, plants or biotopes; or technical systems like laptops or smartphones.

Complexity in a Nutshell

"Maybe I acted too inept when I suggested the theoretical background to my project leader," Priesberg ponders out loud: "What exactly are the essential elements, which a theory for project environments should contain?" Ehrlich replies spontaneously: "I would say 'system', and 'interconnectedness', as well as 'interaction' and ' complexity'. In the future, you will quickly become accustomed to these terms. Actually, it is really easy: First, systems consist of many individual elements, for example the brain contains countless neurons, which are constructed relatively simply, are interconnected to each other and interact via electrical impulses. Therefore, the brain is a system. On top of that: The 'interaction', brings some excitement, since only through

[1] Translated by the authors: "Resist 'process losses' with social techniques".

the interaction of a system's elements does complexity arise: A neuron fires an electrical impulse to other neurons and an emerging pattern of signals has a reciprocal action on the original neuron. By doing this, the neurons influence each other and this can, depending on the kind of interaction, either lead to stable or unstable excited states within the brain, which in turn are responsible for very different behaviors. Thus, the brain can be called a complex system. You can give this example to your project leader. However, I do not believe that he will necessarily be responsive – as he seems trapped by the neuronal patterns which were formed in his own head a long time ago."

As part of the sociological system theory [5], interpersonal communication is also understood as a system. Communication and its related system is considered to be a social "field", with which we, people, interact. With respect to usefulness, the system of communication is only considered to be a sub-system of a social system, which consists of a communication system and its corresponding agents, namely people.

What comprises a system, can be illustrated using a simple example: There are a number of people in a room. At first, there is no (recognizable) interaction between them. Then, an external moderator assigns an interaction activity to the participants: Everyone selects two other people, estimates the distance between themselves and these other two, and then maintains that distance during subsequent changes in position. External intervention by the moderator, who leads one of the participants by the hand to another position within the room, is subsequently followed by a change in position by everyone else. The intervention may result in the system ending up in a stable state after a short period of time or, after a longer period of time, still shows variations, sometimes even with apparently characteristic patterns of movement. If the external intervention is ignored, there will be no cause and effect pattern, but the system as a whole appears to satisfy the rule of interaction. Interactions are obviously characterized by the following fact: Linear cause and effect mechanisms lose their validity, and instead "complex" and reciprocal relatedness of all elements in the system takes their place. What is typical, is that the reaction of the system is independent from the properties of the system's elements. Of course, this is not always true, because if this small experiment is carried out in distinct groups, "deviations" will indeed be observed: Either because someone has asked "why" or because the person affected by the intervention, insists that: "This position is already the one which satisfies the rule of interaction".

A similar example for illustrating the characteristics of a system is introduced using a simple game and its rules by Simon [6]: Every person in a group chooses a three-digit number, the number serves as an address. While throwing a ball to each another, the address needs to be correctly named … To follow the rules, very simple number sequences are used, e.g. 333, 118, 110, 545…after a period of self-organization, certain patterns, which characterize the system as a whole, can be observed. The manifestations of the patterns depend on which numbers are the easiest and quicker to remember and also, which ones are preferred by the group.

As will be explained in detail in the next chapter, the interaction of a system's elements creates a reciprocal "concatenation" of the elements: On the system level, complex overall-behavior is established, which emerges from the interaction on the system's element level. The different overall-behaviors on the system level are connoted as emergent phenomena, which have systemic properties. Systemic properties are properties which cannot be assigned to a single element of the system, but only to the system as a whole.

Projects are, as temporary organizations, social systems with systemic properties which interact with their environment. On closer examination, this environment can again be described as several systems (e.g. the departments of an organization).

In order to answer the central questions raised in the previous chapter "Introduction and Motivation", at first it is necessary to answer the following theoretical questions, so that subsequently, practice relevant questions can be answered. The questions relevant for practice, are marked in italics:

Which interactions occur among the system's elements? How can these be described? *What does that mean for a project team's or organization's action options? Which actions are target-aimed, which ones, by contrast, are pointless and lead nowhere?*

What impact does the system's elements have on the system? What impact does the system have on its elements? *How is the solution to a task influenced by the traits of the project team members? And the other way around: How does the solution influence particular team members?*

Which properties belong to the elements and which ones are systemic properties? *In practice, how can one distinguish the causes from the emerging symptoms?*

How can the interaction of the system and its environment be described? *Which influencing variables can be identified while looking at the cooperation of a project with various other organizations? Which action options arise from this?*

Which overall-behaviors are established by the social system? Is there a pattern and typical parameters which describe systems by their behavior? Which parameters describe the system as a whole? Are there different kinds of parameters and what are they? *What is the essential information required to describe, analyze and understand a project? Which measures lead to success within the social system "project"?*

Which kinds of intervention are possible? Do they differ with respect to impact on the system as a whole or on particular elements? *How can the project team manage to guide itself towards a successful solution?*

Answering these questions takes place in the light of the following basic assumptions:

Standardized methods and best practices are heavily dependent on context and support neither systemic nor holistic perspectives of a complex system. Theories and models depict complex systems and their environments significantly better.

A systemic and integral (holistic) perspective of complex systems will replace organization and leadership, which only focuses on local circumstances ("micromanagement").

People and their needs move into the focus of leadership. Neuroleadership is considered to be the means for bridging human needs and leadership.

Explorative and experimental learning complement learning through the intake of knowledge. Intuition and rationality are equally applied in order to master complexity.

The recognition of (social) interrelations depends substantially on one's ability to view the context and the behavior occurring within the context in an abstract way. For this purpose, meta-competency is required.

2.2 Attentiveness, Mindfulness and Wisdom

> *Polonius: My lord, I will use them according to their desert.*
> *Hamlet: God's bodkin, man, much better. Use every man after his desert, and who shall scape whipping? Use them after your own honour and dignity: the less they deserve, the more merit in your bounty. ...*
>
> William Shakespeare, Hamlet

Theories and models have a threefold relevance for our lives: They are mental images of perceived actuality. They also serve as a basis for the transformation of individual insights into comprehensible and accessible knowledge for everybody. And the third relevant factor is mostly ignored; without a specific "pre-image", we are not capable of perceiving actuality in a conscious manner. The recognition of context and the development of an adequate reaction to a given situation are dependent to a great extent, on our "pre-images". In other words, attentiveness and mindfulness are greatly determined by our ability to make use of mental theories and models for imaging our complex actuality. (For further discussion on our understanding of "theory", we refer to the appendix "Fundamentals Theory and Practice".)

By attentiveness, we do not only mean the level of focus with respect to a particular topic, object or person, but also the ability to recognize the patterns embodied in our environment. Examples of patterns in our communication are e.g. the diverse behaviors of individuals and their relationship with particular motives, values or basic assumptions. Only if this ability of relating things is present, can order be observed in diverse behaviors and the prerequisite for showing mindfulness is be established. Mindfulness means to respect the motives, values and basic assumption of others and to recognize how these motives, values and basic assumptions, together with one's own motives, values and basic assumptions, form a social system of communication on the level of behavior. This is the basis for actively avoiding potential mental blocks and for establishing resonant communication.

With the help of theories and models of personality traits and social interactions, and the willingness and ability to recognize and understand the thoughts, emotions, motives and personality traits of others, our empathy is therefore fostered and in some cases actually made possible. Because, if the behavior of another person is very different from our own, the potential for recognizing and understanding, is for most of us, very slight. In these cases, unjust and negative intentions are easily presumed, which can, on closer consideration, turn out to be positive, yet are tied to inappropriate behavior.

New Methods Require Mental Openness and Systematics

In a quest to conveying theories more easily, on the way to the canteen, Priesberg tells his sparring partner Ehrlich about a colleague, who often drifts off while looking for a solution: "I find this annoying and it's difficult to bring him back to focus." Ehrlich ponders a little, then asks: "When he goes off track, does he subsequently contribute aspects to the solution, which would otherwise have gone unrecognized?" "Yes, this is indeed the case," Priesberg replies. "Then use this property and don't get upset about it," Ehrlich urges and continues: "You will see how theories balance your emotional energy: As soon as you have understood why your colleague acts like he does, you no longer need to be worried anymore and thus have unused energy which you can apply elsewhere in the project."

Priesberg pauses and then carries on somewhat helplessly: "Don't we have to apply and review theories systematically then?" "Smart question," Ehrlich replies. He grins at his colleague and continues: "In order to apply our theories, we need a simple and systematic approach, a framework. I will explain that to you next."

Attentiveness and mindfulness are the core abilities of an empathetic and thus effective executive, and subsequently of an empathetic and effective project manager. One of the aims of this book is to present models and theories of personality traits and social interactions, as tools for cultivating and refining these abilities.

However, these abilities will only develop, if the initial struggle in using the tools is accepted. As soon as sufficient practical experience had been acquired, we were able to observe a substantial impact in the case of project managers:

- (Trained) intuition grows: Complex situations are recognized a lot faster and measures are implemented which lead to success.
- Serenity increases: Perceived self-determination is greater and motives, values and basic assumptions of active individuals are perceived within a particular context. Negative stress only occurs the first time at a much higher stress level.
- Thinking and acting in interrelationships increases: It becomes easier to take a systemic perspective and to identify interactions, instead of linear cause-effect relationships. This is an essential prerequisite to be able to undertake systemic interventions.

These three dimensions: intuition, serenity and the awareness for interrelationships, are the basis for integral comprehension of wisdom [7, 8]. Therefore, one of our aims is to offer tools to interested readers to support them in actively shaping their own personal development.

2.3 The Big Picture

...She takes seriously the butterfly wing that moves in New York and causes an explosion in Bangkok. ...
...It's in Brazil that the butterfly moves its wing, and it causes a hurricane in Texas.'
Does that make any difference, Danglard?

Yes. Because once you get away from the original words, the purest of theories just become rumours. Then we don't know anything. From one approximation to another inaccuracy, the truth unravels and obscurantism takes over.

Fred Vargas, The Ghost Riders of Ordebec,
English by Siân Reynolds

We understand project management as social technology which uses theories, models and methods for shaping and leading temporary organizations.

We assume that within a complex project, which we will further elaborate later on, incomprehensibility and unpredictability will dominate. Therefore, a long-term planning horizon is not adequate for governing actions. Instead, it is necessary to construe the periods of reviewing and adjusting governance as short- and mid-term periods. That is why, we combine the fundamental idea of the PDCA-cycle [9], the success factor-success criterion-model [10] and social techniques corresponding to one specific success factor, to form one operational framework for complex social systems. Figure 2.2 demonstrates this operational framework.

The PDCA-cycle (Plan-Do-Check-Act- cycle), also known as the Deming circle [9] from quality management, was formerly deployed as a model of "continuous improvement" and here is deployed as a core model of "continuous adaptation". Figure 2.3 clarifies the ongoing adjustment of the PDCA-cycle adapted to the respective situation:

For the first step of the "Plan", a project is reviewed at the beginning, or as required according to situation, with respect to success criteria (these are the criteria according to which a project is considered to be successful) and operating success factors (these are the

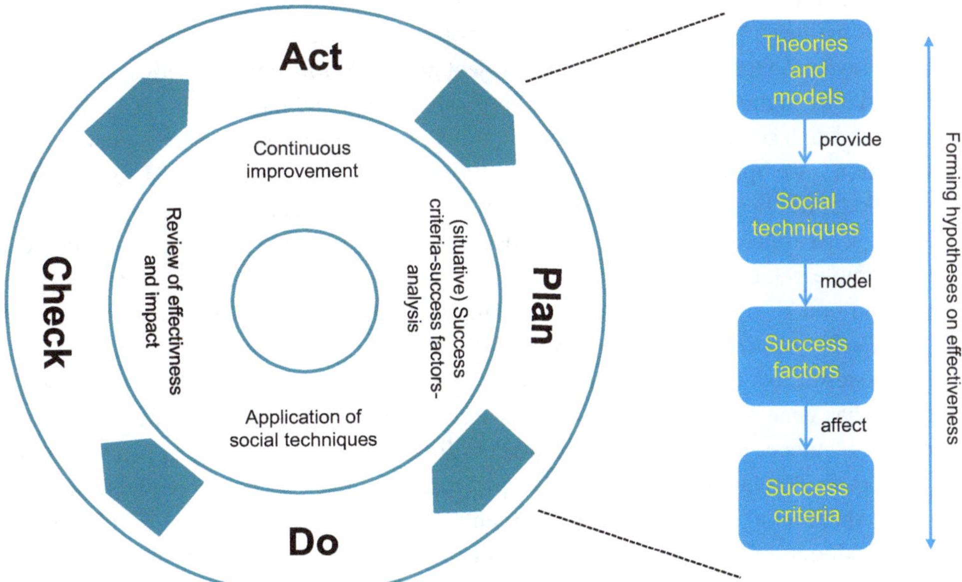

Fig. 2.2 PDCA-cycle, success factors – success criteria, social techniques

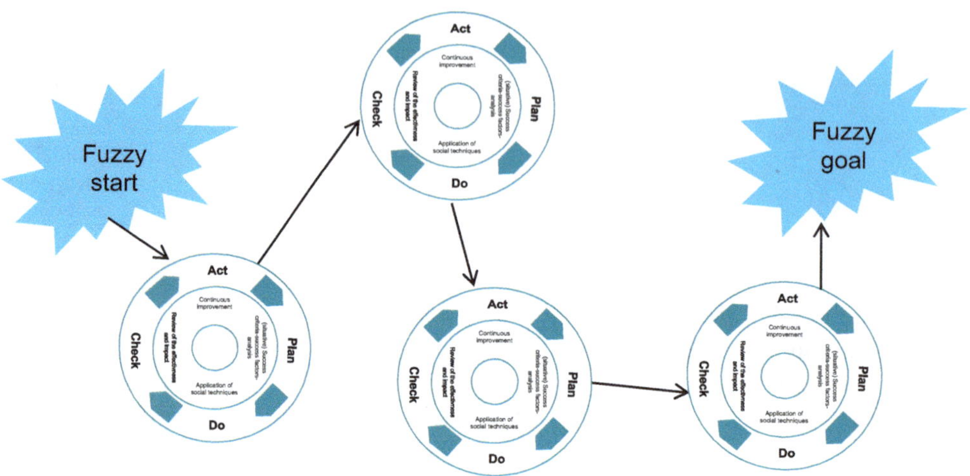

Fig. 2.3 PDCA-cycle and the project pathway

factors which influence the success of a project) [10]. During the course of the project, either the criteria or the factors may change. Additionally, success criteria are not homogenous, because they are dependent on the stakeholders: Each stakeholder uses individual criteria for measuring success. The central message of the model is that success criteria cannot be influenced directly, but have to be influenced by success factors. Therefore, success factors are influence factors for success criteria. For instance, the degree of innovativeness of a product may depend on various influence factors, including team composition. Because of that, team composition could be a deciding factor for whether a product is innovative at all and if so, to what degree and in what manner. The question which can be derived from this is: "What is a team composition supposed to look like in order to develop an innovative product with respect to the project goal?" At this point, theories and models come into play in the form of practically applicable social techniques. Success factors are modeled through social techniques. For example, with respect to the success factor "team composition", we apply the Collective Mind model of team composition. This relatively simple model has been derived from the "Collective Mind Theory".

Most of the time, the theories and models in this book stem from a scientific context and are implemented in an appropriate way for management purposes. Examples are, Kahneman and Tversky's theory of "Thinking fast and slow" or Haken and Schiepek's theory of self-organization. Further on, we will introduce these adapted theories.

In the step "Do", models and theories are applied as social techniques. Once a model or theory has been selected in the previous step "Plan", a particular modification is then used: For instance, a team composition according to the Collective Mind model looks different in an innovative project (Inventor project, see appendix "Fundamentals Diamond Model") than in a simple project (Carpenter project, see appendix "Fundamentals Diamond Model"). Knowledge on the type of project is also very important to be able to make the "team composition" model an applicable social technique. Generally speaking, from this example, we can learn that interaction between success factors implies interaction between social techniques. Further on, we will see, for example, that there is a relationship between team composition, team leadership and the chosen strategy of solution, and even more so, that these relationships may depend on the current situation of the project. In many cases, it is a fact that interdependencies only reveal themselves in practice and elude to a large extent a systematization. This may be proof for the necessary symbiosis of theory and practice.

Once the social techniques of the previous step have been applied, their impact and efficacy should be reviewed in the "Check" step, whilst at the same time their interactions considered. Because of the fact that social techniques are individual models and theories featuring their situationally selected modifications, it could suggest a reductionist approach. The entire set of social techniques, however, is supposed to be seen as a complex network of models for a real project situation, which in turn is supposed to be understood holistically as a "Big Picture". Here it is important to take into account the fact that success factors, as well as the corresponding models and theories and their specifically selected modifications, undergo a steady situational selection procedure. We will see later on that intuition, which we train with the help of social techniques, supports us in this selection procedure. For reasons of validation, we distinguish between impact and efficacy. In an ideal case, the impact, of all social techniques together, can be identified based on the positive development of success criteria. In most cases however, not all success criteria will develop equally and there will be differences in the degree of fulfillment regarding the success criteria of various stakeholders. The abidance of one success criterion, for example the "time and budget" success criterion, can lead to the non-abidance of other success criteria. So, it is possible that for each stakeholder the abidance of the "time and budget" success criterion is associated with a different degree of fulfillment of the "stakeholder satisfaction" success factor. The application of social techniques can also cause delayed negative or positive impacts: For example, a new team composition is only noticeable after a few weeks. Thus, in addition to observed impacts, it is necessary to make an estimation of future impacts, namely of efficacy. This formation of hypotheses on the efficacy of social techniques in particular and as a whole, is crucial for the next step.

Social Techniques: A New Buzzword?

Priesberg reflects on Ehrlich's explanation of some of the aspects of applying social techniques: "All that sounds as if you could not and should not manage any project without this knowledge. Yet, in the past, a lot of projects were finalized successfully without a single mention of the words "social techniques". How can you collate all this?" Ehrlich interrupts him: "Good management decisions intuitively include the correct assignment of people and measures. However, I emphasize again: Our environment changes – experiential knowledge which has proven itself before, may not be applicable in another context. And this is only realized when it is already too late. That is why we should familiarize ourselves with the term 'social techniques' as quickly as possible."

In the "Act" step, the hypotheses of the efficacy of success factors are further tested operationally through interventions within the "project" system. The emphasis here is on small variations to social technologies. The space of potentialities of social techniques spanned by success factors is scanned carefully to find a local optimum in the success criteria. This takes place as long as it is required, or will remain as necessary due to circumstances. This happens so long as the network of success factors and success criteria, including the used social techniques, make a major change necessary. The temporal intervals in which this happens do not follow any regularity, but are subject only to situational circumstances and the assessment of the project leader or the project team.

The model of "continuous adaptation", proposed here was initially applied as part of the so-called Project Navigator [11] and is the basis for a related model developed in the GPM [12, 13]. In the case of the Project Navigator, the review of impact and efficacy is not only geared to measuring success criteria, but also to reviewing whether there appears to be a Collective Mind or not. We will see later on that the Collective Mind is a new collective team state. In the Project Navigator, as well as in the GPM-model, perspectives [14] on success factors are introduced. Depending on project type, different perspectives and also related success factors are important. These perspectives serve to reduce the number of all potential success factors and therefore enable a focus on the important success factors for the project. Seven perspectives (Project Images) have been suggested [14]: Social image, political image, intervention image, value creation image, development image, organizational image and change image. The perspective of transformational management (change image) or the perspective of project political aspects (political image) play a dominant role in a missionary project (high degree of innovation and high degree of mission); in a carpenter project (very low degree of innovation, very low degree of mission) they do not at all (with respect to project types, see appendix "Fundamentals Diamond Model").

2.4 En Route to Complexity

And you see a lot of seaweed all tangled together, making up a sort of Gordian knot. A big heap, maybe a very big one.
Yes
Well, that's what we've got here.
A heap of shit, you mean

Fred Vargas, A Climate of Fear,
English by Siân Reynolds

Let us start our road to complexity by answering the question of how we recognize complexity in a project in everyday life. According to our experience, complexity exists if three kinds of observations are made repeatedly:

- Certain opinions and behaviors of team members are adopted and spread within a network of stakeholders. In some cases, an overall opinion or overall behavior arises

within the network: For example, stakeholders persistently reject a solution suggested by the project core team or accept it enthusiastically. Similar behavior may also arise within a project solution: A change in one element of a solution, e.g. the IT system, has an impact on other sub-systems, which do not have a direct connection to the changed element. In both cases, the social system and the system of solution, interconnectedness plays a central role for the recognition of complexity.
- When small changes have great impact, whether in respect to the social system or the solution system, it starts off by irritating, but then has the potential to immensely increase our sensibility to complexity. In a large number of cases the impacts become apparent much later and almost nobody sees a connection to the original changes: For instance, it is possible that a manager's remark may have long been forgotten unconsciously, but nevertheless leaves a constant imprint on the behavior of employees, even after weeks or months.
- If in addition to the previous observations, and despite profound planning, volatile behavior emerges in the project, then this suggests high complexity: For example, if out of the blue, the behavior of stakeholders changes. In cases where a common goal has already (seemingly) been defined, a whole bouquet of goals suddenly appears. Or the project solution shows system behavior which cannot be explained and does not appear subject to any regularity.

We will see further on that these three observations:

- high degree of interconnectedness,
- small changes with great impacts,
- volatile, non-comprehensible system behavior,

are the central phenomena of complex system behavior.

Complexity: First an Ugly Duckling, then a Beautiful Swan

Priesberg hesitates once again: "I bet most people would call the behavior you describe, chaotic and eventually destructive." Ehrlich interrupts him: "That is exactly why it is so important to carve out the creative force of complexity, in order to lay bare the related potentials."

We come closer to the term complexity by examining it in the light of professional literature.

In their study "Komplexität in Beratungsprojekten"[2], Hanisch and Wald [15] incorporated a distinction of complexity perceived by participants of the study. According to this, there are the following kinds of complexity:

Structural (static) complexity: This is tied to the social structure within the project. This includes interdisciplinary collaboration, a multitude of organizational intersections, and distributed teams, with different cultural backgrounds among other things.

[2] Translated by the authors: "Complexity in consultancy projects".

Complexity of the task: The multitude of subject-specific and technical project content requires several suggestions for the solution with different project contents and hereby generates complexity within the task.

Complexity caused by changes in the course of time: The dynamics within and beyond the project creates a time-referenced complexity within the project.

This classification of complexity constructed in this way can also be found in the evaluation matrix of the GPM[3] for complex projects [16].

Based on this classification, the perceived relationships between the success of the project, or the success of project management, and complexity, are identified by the participants of the study [15]: In the statements of those polled, the statistical analysis of the study suggests a high correlation between structural complexity and the success of project management or success of the project. The complexity of the task and dynamical complexity are significantly influenced by structural complexity. Although according to the study, there is no significant relationship between dynamical and task-related complexity and the success of the project or of project management.

Hanisch and Wald note that "the fundamental assumption for this assertion is that structural complexity negatively influences performance and the ability to coordinate". Thus "it can be deduced that a focus on the reduction of structural complexity is most promising in practice."[4]

We only partially follow this widespread understanding of complexity. We assume an understanding of complexity that follows the Santa Fe Institute's [17] natural scientific understanding of organized complexity and we will show that contained within this, are new insights and consequences for project management.

In accordance with this understanding of organized complexity, complexity emerges in systems through the interconnectedness of a system's elements. These interconnections cause feedbacks among the system's elements, which then evoke non-linear interactions. Associated with this is a high level of sensitivity for stimuli or interventions: Small changes within the stimuli or interventions unleash big changes in complex systems. Cause-effect relationships cannot be identified any longer. So-called emergent macro-states [18, 19, 20, 21] develop, which (often) embrace the system as a whole and go beyond the micro-states of the system's elements. Accordingly, complexity always means dynamic behavior, which means complexity always has a temporal dimension and originates through the interaction of the system's elements. On that note, the "complexity of structures", which could be named static complexity, does not exist. The "complexity of structures" is complicatedness, and at best can be an indication for the likelihood that complex temporal behavior is about to emerge. A short comparison with the observations discussed above, with respect to complexity, shows that these are congruent with the Santa Fe Institute's understanding of complexity.

---

[3] GPM is the German Project Management Association, member of IPMA, International Project Management Association.

[4] Translation by the authors.

Before we illustrate the consequences of this understanding of complexity, we start by introducing Variety. Since Ashby [22], Variety has been associated with complexity and is also often seen as a measure for the complexity of a system. Variety is a number which displays how many states a system can have. On the one hand, the emphasis is on the term "states of a system", and on the other hand, on the form of possibility "can". The discovery that all possible states also contribute to perceived complexity is of enormous significance, because in a decision situation, this is the uncertainty provoking factor.

The term "states of a system", can refer to either the so-called micro-states or the so-called macro-states of a system. We understand micro-states of a system as the states that can be perceived at the level of system elements. Macro-states are those which describe the system as a whole. Because these macro-states denote the properties of a system as a whole and go beyond the particular element of a system, we also refer to this system state as an emergent macro-state. For instance, water is an emergent macro-state, which admittedly is formed by the water molecules, but also goes beyond this. The properties of water (e.g. the macro-state "wet") cannot be derived from the properties of the atoms hydrogen and oxygen, which constitute the water molecule.

In terms of people, our mental states are considered to be emergent macro-states: The interaction of neurons in our brain creates, in a single moment, an emergent macro-state. We refer to the integrated set of all macro-states as our mind. We model the macro-states of our mind by making models of our personality (e.g. temperament model MBTI [10] or the model of basic needs [23]). These models thus depict one part from a variety of macro-states of our personality. The modeling of mental macro-variety certainly does not describe by any means, the whole personality of a person. Yet, experience shows that doing it provides comprehensible access to the complex human world. However, it is – as we shall explain below, still rarely possible to obtain by modeling recommendations, with which the future behavior of a complex system can be deterministically predicted or even influenced.

Characteristics of organizations, and indeed of society itself, are also regarded as emergent macro-states. They are created by us, through communication. Communication is interconnectedness, which give rise to non-linear, mental interactions. Again, we try to approximately depict the variety in the "organization" system through models of organizational macro-states [24, 25].

However, it is not possible to determine the variety within a system by counting system elements and identifying their structural links, because system states emerge through the interaction of system elements. Counting individuals or organizations and the number of related links is not a measure of the variety or complexity of a system "project", on either a micro-tier or a macro-tier. This is a commonly existing misconception, which also means that so-called simple systems, consisting of only a few elements, are not attributed with complex behavior. At this point, the distinction between complicated and complex systems is deliberately introduced. Complicated systems are difficult to understand, because they contain many elements and links, but their interactions do not evoke new system quality. In this respect, it is often asserted that only technical systems are complicated and only natural systems are complex [26]. This distinction is not made in complexity research [27, 21] and occasionally can even lead to very dangerous mental bias (we refer to the section further on "Interplay of Intuition and Rationality"). Dörner [28] elaborates in an impressive way by using the Chernobyl disaster as an example, a devastating interaction of technical and social complexity. He clarifies in particular how ignorance of the complexity of the nuclear reactor, enhanced by various mental distortions, provoked a chain of situations with a catastrophic outcome. It is all the more tragic, when, even nowadays, the myth of only technology being complicated is still repeated in stereotypical fashion. In the 1970s, complexity and chaos research had to expend a great deal of effort to overcome corresponding belief systems. It is only recently that Appelo has presented, as a first in the area of management, a more nuanced view for Agile Management [29].

On the basis of this explanation, we derive the following. In a project there are two systems which feature complexity: the social system and the intended project solution as a technical system. Whereas the task (the scope), the social structure (stakeholder and social organization) and the context (environment) are drivers of complexity. We emphasize this difference because, as we have seen above, from this, several misconceptions originate regarding complexity.

We understand complexity drivers as factors which facilitate interconnectedness and thus generate complexity via non-linear interactions. Complexity drivers affect each other reciprocally and create social complexity within a social system and solution complexity within a solution. Figure 2.4 illustrates these relations:

Let us have a closer look at complexity drivers:

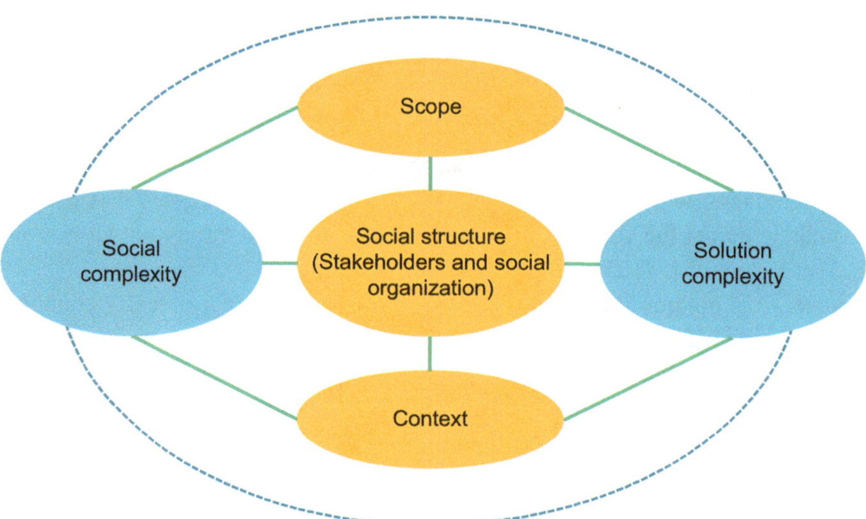

Fig. 2.4 Complexity drivers (in the middle) und complexity domains (right and left)

Complexity Driver Task (Scope)

Based on our experience, the complexity driver task (scope) can be described very well using project typology with respect to the diamond model [30, 10, 31]. In the diamond model, the project task can be characterized according to four dimensions: Degree of innovation (Technology), degree of mission (Novelty), degree of management (Pace), and degree of abstraction (Complexity) (also see appendix "Fundamentals Diamond Model", where we use dimension denotations from [10] and from [31] in brackets).

The degree of abstraction (Complexity) indicates the extent to which the task creates a solution consisting of systems of systems. Here, for example, from the complexity driver "degree of abstraction of the task", we can extrapolate to a solution's correspondingly complex (dynamic) system behavior. The Berlin underground network, the network of the "Deutsche Bahn" or highway electronic toll collections, have a high degree of abstraction, because they are systems of systems. Under certain circumstances these systems show complex dynamics. In the realistic thriller "Black Out" [32], in which interactions of technical and social complexity are described during the collapse of the European electricity grid, this perception of complexity can be understood.

The degree of innovation (Technology) indicates to what extent the project team need to develop new natural scientific or social techniques, in order to define the goal and get closer to it. In addition, this dimension evokes uncertainty among stakeholders, and therefore social complexity. In the project solution itself, unpredicted interactions may be caused by new technologies, which then create solution complexity.

The degree of mission (Novelty) takes into account, the extent to which all stakeholders are familiar with the task, and also specifically, the intended solution. A solution, which for example, requires new operational processes, is very likely to cause emotional barriers at first. In this case, the solution drives the complexity, even though in itself, it is not complex at all within the social system, yet causes social complexity.

The degree of management (Pace) measures to what extent stringent leadership is required due to the task. The autarky of individuals or teams can be considerably reduced in this case. Examples of these are crises: Due to the high degree of management, teams lose much of their self-determination, which may be contrary to basic human needs. This increases the hazard of social complexity.

We use the diamond model to create transparency within the team for the complexity driver task (scope). In an interdisciplinary team consisting of, for example, mechanical engineers, hardware developers and software developers, it is extremely important that all team members have the same understanding with respect to degree of innovation in the project. If, for example, software developers are unaware of the innovations required in the hardware to make their software "run", this could result in high friction losses. It creates value-destroying social complexity. Very often, this is also associated with value-destroying complexity in a technical system.

Complexity: Always Value-Creating?

Ehrlich does not tire of picking the brains of Priesberg's project leader: "You see, this man creates value-destroying complexity by labelling you as theorists and therefore undermines your professional authority. As a result, you are blocked, instead of being able to deliver your full contribution to the team, and subsequently, your team is unable do this either. You cannot succeed in aligning our team members as long as the control parameter set by the project leader continues to work and induces value-destroying complexity. Therefore, it is so important to understand and to master the value-creating and value-destroying facet of complexity."

We speak of value-destroying social complexity in an organization, when the value creation of an organization is reduced by structures and processes or by interest and power struggles. In this case, the actual value is below the potential value of the organization. In a worst case scenario, it could be that "the whole is less than the sum of its parts". On the contrary, value-creating social complexity exists when interconnections within a team or within the organization lead to the fact that the whole is more than the sum of individual contributions of team members. For example, a high-performance team delivers a value, which goes well beyond the individual contributions of team members. Below, we will see that a high-performance team is formed by mental interconnection. Even a task itself can result in value-creating or value-destroying complexity of a solution. We speak of value-creating complexity of a solution, if a project team can develop a complex solution, which contributes positively to the value creation of a company or organization.

Complexity Driver Stakeholder and Social Organization

With their personalities, interests and spheres of influence, stakeholders are, per se, a source of complexity. Personality preferences, motives, values and belief systems, are mostly in the background and unconsciously effective as distortions (see the section "Interplay of Intuition and Rationality"). They can cause unforeseen forces, and through the interconnectedness of communicational processes, have positive, as well as negative consequences. The underlying organization with its processes, structures and cultural elements (i.e. values, belief systems, stereotypical behaviors) can further increase complexity. The task of an organization is mainly to regulate and absorb this complexity. In the perception of affected members of an organization, however, complexity is often neither absorbed, nor reduced, but created. We all know cases where complexity grows, due to stakeholder's specific likes or dislikes, or due to traditional moral concepts in communication and behavior, and where often, needless complex project solutions are produced. In such cases, social complexity affects the solution negatively, thus creating value-destroying solution complexity.

However, there are also cases in which social complexity is value-creating. Further on, we will continue to pursue the question of what organizations should look like, in order to function and regulate and absorb with respect to complexity. Because of their enormous and underestimated importance, we have devoted an entire section to the positive impacts of high social complexity (see the section "Regulation of Complexity Through Targeted Interconnectedness and Self-Organization").

Complexity Driver Environment (Context)
The environment of a project (for instance other projects, its own enterprise, markets, partners, customers, society, politics, as well as elementary risks) has its own complexity. From the perspective of a project, this complexity acts rather like "random" events. Changes in the environment act as complexity drivers via the social system and the task. Obviously, it is of great importance to take into account the complexity driver environment as potential uncertainty and the associated risks in the social system and in designing the project solution. In this book, however, we do not elaborate further on this complexity driver, because on the one hand, it is too multifaceted, and on the other hand, within a project, its implications are reflected in the other two complexity drivers.

All three complexity drivers, i.e. task (scope), stakeholders and social organization, and environment (context), cause an interconnectedness of social and technical systems with the respective complex dynamics in the social-technical realm.

2.5 Value-Creating and Value-Destroying Complexity

Watch it, little Marie-France, watch it. … Turn your thought over seven times, but not ten, or twenty, her father would say, otherwise it gets worn out and you'll never find the answer. There are people like that, who go on thinking in circles for ever, it's sad, just look at your uncle.

Fred Vargas, A Climate of Fear,
English by Siân Reynolds

Complexity research [33, 34, 35, 21, 36] has shown that emergent macro-structures can be described fairly easily, provided we familiarize ourselves with the associated range of thinking tools.

The start of complexity research in the 1970s was characterized, in particular, by the significant challenges posed by this otherness [27]. We summarize this otherness as follows:

- Even simple systems can exhibit complex behavior.
- Potential complex behavior of systems cannot be recognized by inspection. Natural, social and technical systems may exhibit complex behavior.
- A (mathematical) description of complex behavior can be surprisingly easy.

Figure 2.5 shows that (often only) two types of parameters are of crucial importance for describing organized complexity (we shall see below, that there is an additional type, setting parameters): These are the so-called order parameters and control parameters.

We refer to our article [38] below.

Control parameters are parameters, which, during variation, indirectly have the ability to influence the formation of macro-states[5]. In other words, they affect a system's ability of

[5]Think of a cooking pot filled with water, in which the temperature difference between the lid and the base is the control parameter.

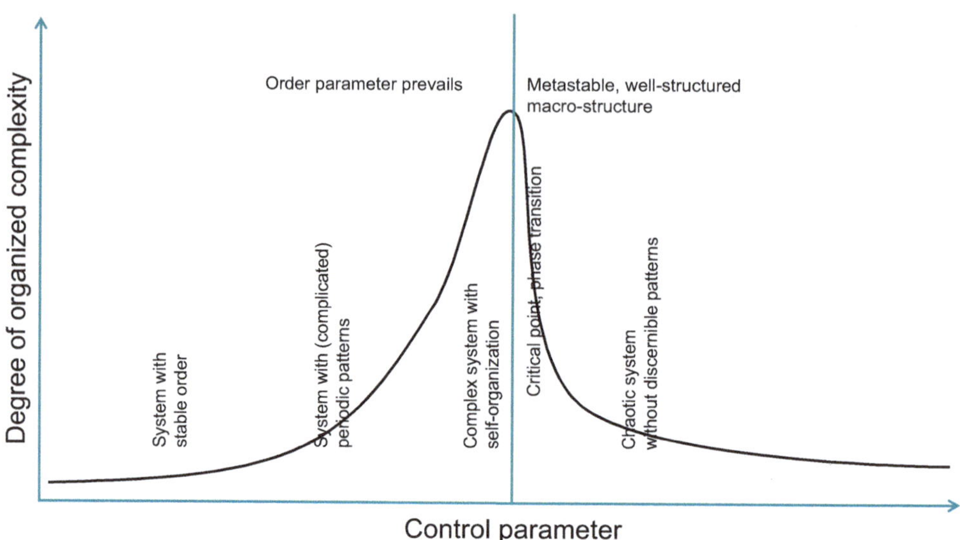

Fig. 2.5 Degree of organized complexity in systems, with reference to [37] and using Wolfram's classifications [21]

self-organizing macro-states. Control parameters may be internal system parameters, or external stimuli, which act upon the system. We have already become familiar with complexity drivers: Below, we will see that certain complexity drivers can function as control parameters. It is one of the responsibilities of project management, to identify control parameters and deploy them purposefully for managing a project.

Order parameters are parameters which, shortly before transition to a so-called chaotic system dynamic, describe and induce a dominant order[6], namely an emergent macro-state. In mathematics and the natural sciences, emergent macro-states are referred to as attractors. Hence, they are macroscopic states which "attract" the whole system. The term order parameter accounts for this "organizing" characteristic: It is a parameter, which induces order within the whole system. This, however, is not a stable order, but a dynamic one, which can be altered or destroyed by the smallest stimuli or interventions. Therefore, the system reacts adaptively to stimuli or interventions.

Along the control parameter, Fig. 2.5 shows four stages:

- Systems with a stable order,
- Systems with (complicated) periodic patterns,
- Complex systems exhibiting self-organization,
- Chaotic system without identifiable patterns.

[6] If the temperature difference is sufficiently large, thermal convection emerges in the cooking pot, and thus a macroscopic flow pattern. The macroscopic flow pattern (stable convection current) is the order parameter and "forces" all flow particles to follow this pattern. Thereby, a pattern is enforced.

These four stages are the so-called Wolfram classes of complexity [21] (see appendix "Fundamentals Complexity Classes"). As you can see, the ability of a system to form organized complexity increases with the variation of the control parameter; and declines abruptly again at a "critical point". The critical point is the transition from complex behavior to chaotic behavior.

Only complex systems at the edge of chaos show an ability to form emergent macro-structures. Although to a lesser extent, chaotic systems show macro-structures, no dynamic order patterns can be detected here. Systems with a stable order, like the name already suggests, form macro-structures, however, any alteration through external stimuli or interventions always results in the same macro-structure again. Systems with periodic patterns exhibit a little more variety, but literally go around in circles. As a result of their sensitivity to external stimuli and interventions, complex systems are capable of taking in information in various ways and of adapting accordingly (see the section "Philosophy of Complexity"): Complex systems form a variety of macro-structures. From Fig. 2.5 another central statement can be derived: A system can, depending on the choice of control parameters, show either a stable order, an order with periodic patterns, organized complexity or chaotic behavior. Therefore, complexity is one of the behaviors that can be shown in the same system under specific conditions.

These four complexity classes, experienced, albeit in a simplified form and not with the intention of classifying complexity, wide distribution via Snowden's Cynefin Framework [39]. Here, the following labels were introduced for the four domains: simple, complicated, complex and chaotic. Unfortunately, this simplified form led to various misunderstandings: Among other things, that (simple) systems with only a few system elements cannot exhibit complex behavior and that (complicated) technical systems also do not show complex behavior.

Systems at the edge of chaos show, in their new dynamic order structures, "simplicity". Because of these order structures, the order parameters, act in an "enslaving" way upon the individual elements of a system and give all members a certain "alignment" [34]. This also means that even systems with many elements can be described at the edge of chaos using a few control parameters and order parameters. In this sense, complexity works in an inclusive and simplifying way! Table 2.1 lists examples of emergent macro-structures, as well as the respective control and order parameters. Here we use the terms "control and order parameters" as technical terms, even when in a social context this choice of words may appear strange to some readers:

The task of project management, therefore, is to operate in a complex project at the edge of chaos and to use the respective variety of emergent macro-structures in a value-creating manner. It is the job of the project manager to identify in the above-mentioned complexity drivers, the control and order parameters for social complexity and solution complexity, and subsequently apply them within the project. Hence, we also refer to the ensemble of these systems parameters as leadership parameters. Within the following chapters, we will look more closely at this task.

Table 2.1 Examples of complex structures, their control and order parameters and the emergent macro-structures

Example (natural, technical, social, socio-technical)	Control parameter	Order parameter	Emergent macro-structure
Rayleigh–Bénard Instability (natural)	Energy input	Dominant fluid dynamics	"Fluid" honeycomb pattern
LASER (technical)	Stimuli light (energy)	Dominant light wave	Coherent light
Society (social)	Identification, emotional safety	Values, belief systems	Culture, form of state
High performance team (socio-technical)	Balance of requirements and abilities per team member (Neuroleadership)	Construction of meaning via integrated information by use of Collective Mind scheme	Collective Mind
Critical Chain Multi-Project Management (CC MPM, social-technical)	Work-in-progress of the MPM system	Fever curve consisting of project progress and buffer consumption	Flow in the MPM system

The Goal: The Emergent Macro-State

"Wait!", Priesberg thinks out loudly and repeats: "Project leaders are supposed to throw projects into chaos so that *what* happens?" He pauses. Ehrlich has anticipated this reaction and interrupts him: "You have heard quite right: Project management should lead a project into the domain in which the formation of emergent macro-structures, which can be described through simple order parameters, is most likely. Only then, in a complex environment, can a project be regulated, and only through simple models. Therefore, we do not speak of Project Management at the Edge of Chaos, without good reason."

Priesberg replies provocatively: "But a question instantly arises in my mind: Does my project have the ability at all, to form these emergent macro-structures as you call them? And, how do I know that? And, what do I do if there are no emergent macro-states?" "It is exactly these questions that we are going to investigate very thoroughly, and in order to this, we next need to study complexity very carefully," Ehrlich concludes seriously.

2.6 Complexity, Suspense and Uncertainty

Go make thyself like a nymph o' th' sea:
Be subject to
No sight but thine and mine, invisible
To every eyeball else. Go take this shape,
And hither come in 't: go: hence
With diligence. ...

William Shakespeare, The Tempest

Complexity, suspense and uncertainty are words, which are often used synonymously. Often, complexity is understood as a subjective feeling: "But that is complex". What is actually meant is that in this case, you do not have an overview of the situation or the problem, or that you expect the unexpected.

In the previous section, we have seen that it is useful to distinguish between complexity drivers and complexity domains. Complexity, as we understand the concept, is a property that we assign to a system. Whereas terms like suspense and uncertainty are concepts in which an impact which affects us is expressed. This is why, in this section, we will look more closely at the concept chain of complexity drivers, complexity, suspense and uncertainty.

Above, we became acquainted with project tasks, stakeholders, social organization and the environment, as complexity drivers. Complexity, considered as a property of a system, is distinguished according to the domains of social complexity and complexity of a solution or a product.

With respect to the consequences of complexity, initially we used the terms incomprehensibility and unpredictability. Incomprehensibility underlies the fact that we cannot capture a system, either in detail, or as a whole, at any given point in time. The term unpredictability, additionally brings the time dimension into consideration and expresses the fact that the future holds unknown facts beyond the norm.

In the same lexical field, the terms risk, suspense and uncertainty are used. Usually, no neat distinction is made between these words. Gigerenzer [40], points out the difference between risk and suspense. In decision theory [41], the terms are used in a similar way to how we will use them.

We define the terms risk, suspense and uncertainty using the two dimensions, degree of mission (Novelty) and degree of abstraction (Complexity) of the diamond model (see appendix "Fundamentals Diamond Model").

Figure 2.6 clarifies, based on the two dimensions of the diamond model, the relationship of these terms with one another. In addition, we have used the classifications (known knowns, unknown knowns, known unknowns, unknown unknowns) popularized by the U.S. Pentagon [42]. The assignment to these classes can be illustrated by the following simple example: Imagine, you are listening to a lecture, and the speaker holds his left hand behind his back. You do not have the slightest idea what is driving him to do that. You may suspect that he has never received any training in how to hold a lecture, or that maybe his hand hurts, or his hand is dirty and he wants to conceal it. In short, you do not know what you do not know. It is the area of highest uncertainty (unknown unknowns). If the speaker moves his hand forward and opens it, you will see that he is holding a coin. You now know the object (a coin), however you do not know what kind of coin (e.g. 1 Euro coin), or why he is holding this coin and what he is attempting to do with it. You are in the area of unknown knowns, or of suspense. While you know that the object exists, you do not know anything about its properties.

When the speaker tosses the coin, you already know quite a lot about the "characteristics" of the object "coin". In addition, as it falls to the ground, you can specify the probability with which the coin will display heads or tails, before it hits the ground. The moment

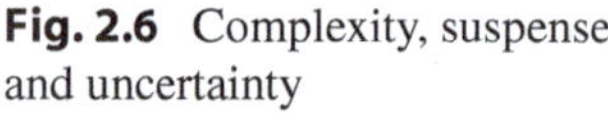

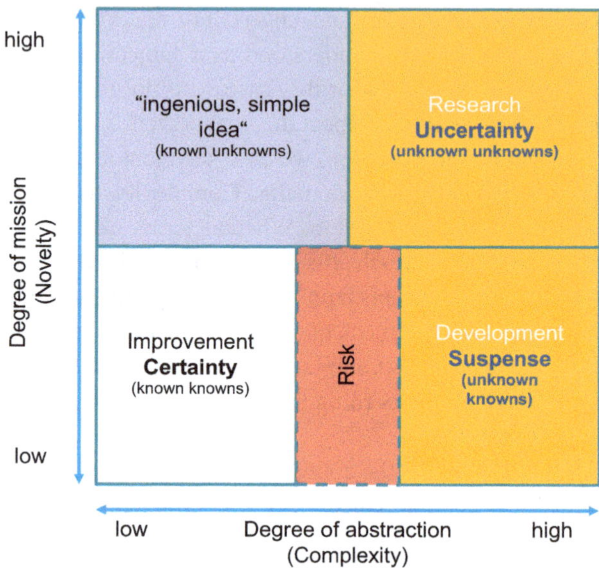

Fig. 2.6 Complexity, suspense and uncertainty

when the coin comes to rest on the ground, you abruptly gain certainty over the coin and its property "heads up" or "tails up". Only in the area between unknown knowns and the area of known knowns, are you in the area of risk: You know the potential properties of a known object and you can make statements on the probabilities with which one of these known properties will be realized in actuality.

The domain of "known unknowns" is an ingenious, yet simple idea: objects or ideas are used in a simple, but new way or are connected to other objects or ideas. A prime example is the "Post-it": The combination of a slip of paper, adhesive and an application domain, created a new and simple product.

Why is it important to become aware of these different domains of incomprehensibility and unpredictability in project management?

The domain of risk is one in which all objects are known and associated events can be calculated in terms of their probability of occurrence. Incomprehensibility and unpredictability are present, but computable.

The area of suspense is one in which all objects are known, however, associated properties and events in terms of quantity and quality of their occurrence cannot be estimated. Incomprehensibility reaches a new level: There is something about the objects, which is completely unknown, and whether and when they occurs, is also completely unknown. This is the domain that Nassim Taleb refers to as that of black swans [43]. While we are aware of the existence of swans, we have no idea in what actual color they can still occur. This area is therefore associated with the (further) development of existing products or development projects.

The domain of uncertainty is one in which all objects are not yet known and therefore nothing can be said about associated events. The area of uncertainty is one of maximum incomprehensibility and unpredictability. It is not even known that there are swans, so the properties they have cannot be known either. We therefore associate this area with research and innovative projects.

The domains risk, suspense and uncertainty function to different extents as complexity drivers: The probability of induced social complexity and solution complexity increases significantly along these domains.

Intuition is often regarded as the weapon of choice to master complexity. The classification mentioned here refers to different forms of intuition, and specifically to the fact that unconditional application in all domains, as a means to cope with complexity, can cause serious fallacies: In the world of suspense and uncertainty, our tendency to mental distortion (bias) increases significantly. We will examine this later in the section "Interplay of Intuition and Rationality".

2.7 Philosophy of Complexity

His speech was like a tangled chain; nothing impaired, but all disordered. ...

William Shakespeare, A Midsummer Night's Dream

After addressing the manifestations and terminology of complexity in the previous section, in this section we will illuminate complexity in the light of system theory [44, 5, 1, 45, 46]. In this way, we will have a mental warehouse for dealing with the regulation of complex systems in the next chapter.

In the section "From Negative Words and Basic Assumptions" we have explained the term "system" using as an example, a group whose members are "connected" via a spacing rule. Intervention into the system in the form of "a person being taken by the hand and guided to another place" resulted in a change in the overall arrangement of the system. Now imagine for a moment, the spacing rule was lifted: So there is no longer any "connection" between people . Now, as before, you still want to bring everyone into the same desired target state position. What would the effect be? Since no one is connected anymore, each person would have to be taken individually by the hand, to bring them to their respective new destination. The consequence would be, a lot of work.

In other words, the spacing rule, which generated a connection between individuals, saved a lot of work. Connections, a central feature of complexity can create value: Here, connections save intervention energy.

Now we can add another rule to the spacing rule: "Every person chooses a point in the room and determines their viewing angle towards this point. This viewing angle should be kept while changing position". What effect does this have? An intervention from outside, in the form of "a person being taken by the hand and led to another place", should normally lead to a system reacting more inflexibly. There are more connections, but they may not provide any value. The system freezes due to the connections, and potential dynamics are lost. Too many connections require additional intervention energy, as do too few connections or no connections at all.

This observation is the basis for the frequently encountered negative connotation of complexity. There are many connections in a company in the form of standardized processes, rules or roles, but they are not perceived as value-creating.

Connections are important, but they alone do not produce complexity. This only exists when the system has just the right number of connections to remain dynamic. In this case, it is able to take in information. In our example, the system of people connected by the spacing rule, has adopted information, which is a result of an intervention.

Based on this small system example, we can illustrate something further: We can lift all the rules and replace them with the statement, "Whatever happens in the future, please act as a team." At first glance, this statement could again, be interpreted as a rule. At second glance, however, it is a statement on the value of "team orientation". The intervention "a person is taken by the hand and led to another place", in this case, also results in the fact (to date, it has always worked during training by one of the authors) that a group of people move to a particular place determined by the intervention. In contrast to the application of the rule, there was no attempt to preserve distances. Therefore, much greater differences to the initial configuration were clearly visible, during the changing of position and thereafter at the final position. The value "team orientation" leads to the fact that when everyone in a team has a similar understanding, the group stays together as a group. The value "team orientation" acts as an order parameter. This order is aligned by the value and leaves more options open to behavior. Intervention has an effect, but the associated new order has been carried out with much more flexibility.

In the examples given in the next chapters we will learn that a value such as "team orientation" can have different effects: A value can occur as a control or order parameter for example. In the above example, the common perception with respect to "team orientation" value was used in the group, to achieve cohesion, or an "aligning" team spirit within the group. For this reason, we also speak of an order parameter. A control parameter "team orientation" exists when individual team members arrange common tasks, decisions and procedures for example, and perform as required. In this case, team orientation is a kind of prerequisite so that an order parameter can unfold, for example, to jointly develop a target and pursue it. One can then also recognize that parameters can occur in hierarchies. If the team itself specifies values alongside goals, we speak of an order parameter hierarchy. If a team gives itself the value "team orientation", while at the same time pursuing a common goal, an order parameter hierarchy develops consisting of goal and value "team orientation". In general, a parameter hierarchy exists in complex social systems for all parameters.

In the chapter, "Consequences for Management Systems" we will return to our above example of a system.

Emergence Only with the Correct Order Parameters!

"Here we have a very good example of an order parameter," explains Ehrlich and continues: "If you, the project leader, worked out with the group what constitutes a team and how it can react in a solution-oriented way, then this single switch can ensure constructive collaboration takes place."

Priesberg interrupts him: "How does that work? Should I just say 'abracadabra'?"

Ehrlich laughs and replies: "Yes, of course" and adds: "And if you know that a group has no team spirit and you still demand it, then the project can fail and why is that?", he continues, talking more to himself: "Because everyone in this unformed group understands something different by team work. Therefore there is no focus on a solution and the project is doomed to fail."

"So how does this 'abracadabra' order parameter work?" asks Priesberg impatiently.

"It's simple: If you know exactly how your team has worked in the past, perhaps even with flow experience, then a request to work as a team in the future, also supports the search for solutions. This can be very helpful, for example in uncertain situations. However, if you have a group in front of you which has never worked as a team before and you still request teamwork, you do not allow the group the necessary time required to form a team. Then everyone works in their own way on the project, which can lead to either a suboptimal solution or no solution at all. Managers should therefore consciously and carefully encourage the formation of an order parameter, which gives a direction to the team."

The one organization that creates the most value by far, is nature. The extraordinary thing here is that, at least as far as we know, no one on the outside determines what is valuable and what is not. Nature discovers this for itself and this finding is usually referred to as evolution. Nature constantly builds new connections, "rates" these connections and, if the "value is not recognized", deconstructs them again. This recognition is not a "conscious" recognition, but is "the difference that makes a difference". Therefore, nature selects the "right" states, to keep a system, i.e. an organism, alive. This is one of the most well-known formulations of systems theorist Gregory Bateson [47], which is also often alluded to within the context of Niklas Luhmann's social system theory [5].

While Bateson intended to be understood in accordance with the universal title of his book "Mind and Nature, a Necessary Unity", Luhmann applies this formulation mostly to social systems. The statement "the difference that makes a difference" is closely linked to the concept of self-referentiality. By referencing to itself, the system perceives a difference, which makes a difference to the system. With the mechanism of self-referentiality and the related selection of "right" states, the system keeps itself alive. This is only valid as long as environmental conditions do not change. If that happens, and especially if it happens too fast, the system cannot adapt – it dies. We just need to consider, the oft-cited example of the extinction of the dinosaurs due to a single event (meteorite impact).

While the concept of self-referentiality in social system theory is reserved for living systems, we use self-referentiality for all systems, inanimate as well as animate, namely as an ability to perceive a difference that makes a difference, and to cultivate this difference in reference to itself, i.e. the system in distinction to its environment. Therefore, it is not the conscious perception of this difference which is essential for us, but only the "self-referential" relationship to the system as a whole.

The Difference that Makes a Difference

Priesberg looks irritated when he meets up with Ehrlich in the canteen, to explore the depths of complexity: "'The difference that makes a difference', what is that supposed to mean again?"

Ehrlich leans back relaxed and replies enthusiastically: "Systems that keep themselves alive, so in a stable state, need to constantly select things from their environment. The system selects the 'right' things, it notes the differences between the things available for selection. It is precisely this distinction that keeps the system alive and that makes a difference. And the great thing is: The system does this quite unconsciously".

"Peter Piper picked a peck of pickled peppers…," Priesberg mimics Ehrlich's rapid speech. "It couldn't be more abstract, could it?"

"It is quite simple", Ehrlich continued unaffected by the mockery, "just now…., you are eating your soup. Thereby you are keeping yourself alive, because nutrition, and therefore energy, is absorbed from the environment. And because it is tasty, we can assume that it is not something which would compromise the maintenance of your emergent macro-state, namely your body. So, you distinguish naturally and take in the right things from your environment, in this case nourishment. By these means you continue to remain with us, you who obviously distinguish yourself from your environment. Therefore, this is a difference which makes a difference."

Priesberg places the spoon audibly to one side: "Our department for regulatory affairs comes to mind. Try pushing through something new. It won't work. Neither by dropping a bombshell, nor using a gentle whisper. 'Mr. Priesberg, we are not interested in your ideas,' I hear constantly."

Ehrlich added hastily: "Even organizations as social systems represent an emergent macro-state that they must maintain through the interaction with the environment. In the case of organizations, however, it is not food that is crucial, but communication, which is selected from the environment and brought inside. If a new idea is introduced from the outside, the organization automatically decides how to handle it. In your example of a rigid organization, the idea is simply rejected."

"How long can this continue like this?", Priesberg asks himself.

"As long as the products or outputs of this organization are required by the environment, everything is fine, but once they are no longer required, the organization crumbles. And that happens much faster in organizations resistant to change than in flexible ones, which are open for change. For the latter, at least, there is a chance to embrace change and any corresponding reaction to the then unfamiliar type of environment. And when an environment changes slowly, then an open and adaptable organization can most definitely survive," Ehrlich adds.

Priesberg scratches his head. "At the heart of the department for regulatory affairs are three IT systems. All requests from customers, for example requests for the registration of new active ingredients, are described in the language of the input variables of the IT systems, and everything which comes from the outside, from customers, is mapped with these variables. It even gets to the point where employees think in the logic of these IT systems."

"This section, for example, forms an emergent macro-state. And as long as it is needed, it will remain alive and only receive communication from the outside, which it can process with the help of the IT systems. This is also true for the personnel: It is, and behaves structurally identical. People who differ substantially in terms of temperament, values or basic assumptions are usually not even accepted in this department, or if they are, there is a high probability that they will leave after a while. This greatly stabilizes the emergent macro-state of the department. Let us hope that their environment remains stable for a long time, otherwise …" Ehrlich grins.

Macro-structures in a complex system can only be formed if such self-referentiality exists: Each system element perceives its neighborhood within the system and orients itself according to this perception. This is also the reason why, in complex systems, simple cause-effect relationships no longer exist between objects that act as causes for other objects: All system elements perceive the system at a particular point in time and align themselves accordingly. This happens continually by forming new macro-structures. We talk about macro-structures, when a visually perceptible structure forms, or more generally of macro-states[7], if we mean all macro properties (including structure) of a system.

The so-called Wolfram classes of complexity can be clearly understood on this basis, (see appendix "Fundamentals Complexity Classes"): According to these, there are simple systems in which, after a short time and no matter with which initial state they started, the same final state is always reached. There are complicated systems, which are systems that, regardless of which initial state they started off in, periodically go through a number of different states, but then no further. There are complex systems that for different new initial states, always form different new macro-states. These macro-states are not stable, but fragile: Small interventions lead to new fragile macro-structures. Complex systems are thus able to represent highly diverse information: Each initial condition or intervention leads to another system macro-state, which is nothing more than a representation, or we could say, model of the intervention. Furthermore, there are chaotic systems; systems that form, according to the initial state or the intervention, new macro-states. However not one fragile macro-state is formed, but a whole flood of states in the form of state sequences, which also exhibit no regularity. Chaotic systems are not capable of representing information, as stable and complicated systems are. Chaotic systems are systems in which there is a network of system elements, but the context conditions (setting parameters), and especially the control parameters, do not allow fragile macro-states to form. Very often events are referred to in colloquial language as chaotic, if they are perceived as coincidences: A system will also be perceived as chaotic if many "external" events or hazards, from the environment, act upon the system and disturb it.

Therefore, system setting parameters and control parameters may change due to "external" events and subsequently no order parameter is able to form. The system then behaves as a stable, complex or chaotic system, depending on the parameters. The term 'disorder' refers to the absence of order and is not the same as chaos, even if it is commonly used

[7] State is a word that is often associated with statics. We do not understand state in this sense, but adhere to the term from physics, where it is understood as a dynamic collection of properties.

interchangeably. In disorder, there are no links between elements of a collection of elements. We no longer talk of systems, because individual recognizable elements are unrelated and perceived as completely arbitrary at one specific time in one specific place. Accumulations with unrelated elements are, as we have seen in the above example of the individuals subject to the no spacing rules, able to influence as a whole using a great deal of energy.

We emphasize that a system's (or an object's) properties, such as "simple", "complicated" or "complex", always become apparent in a certain context. So, there is no "simple system", but a system is simple in a specific context. This is one of the key reasons why complexity research had a very hard time at the beginning: A "simple" iterative equation from which a sequence of numbers is generated, generates complex patterns within the numbers, a simple mechanical system like a double pendulum exhibits, in a particular context, complex movement patterns in space and time.

This means that complexity and simplicity are not opposites. Therefore, when we talk of a simple, complicated, complex or chaotic system, we mean that the system displays simple, complicated, complex or chaotic behavior within a specific context. In many cases, it may be that a system that is referred to as simple, shows simple behavior in several contexts.

A system is perceived as simple if the interventions into a system are of an "easy" nature: In the above example of intervention, in the group of people, the intervention is perceived as easy if it shows an effect without much work. In this case, however, it is precisely the complexity of the group (i.e. here its interconnectedness), which is a prerequisite for this simplicity. Here again complexity and simplicity are not contradictions.

The number of system objects is therefore not a prerequisite for complex behavior. One, two, or more individuals, can demonstrate complex behavior. The same applies in principle to any other type of object. In principle means, in this case, complex behavior of one, two or more objects, cannot be excluded per se, or that there is currently no verifiable statement, according to which complexity is excluded for systems consisting of a few objects.

We often intuitively answer the question "Does our world become increasingly complex?", in the affirmative. The interconnectedness of our social life is steadily increasing. Financial activities in the USA and China have an impact on financial markets in Europe, the emission of greenhouse gases contributes to global climate change and is also felt in regions such as Antarctica that contribute little to greenhouse gases; or in social networks, networks emerge which suddenly erupt into mobbing. Networking, high sensitivity with respect to initial conditions, suddenly appearing new macro-states (changes in financial markets, and mobbing can promptly turn into chaos.

In the above examples, the criteria for complexity have been fulfilled according to our understanding, so our social complexity increases, and the complexity in our technological system (traffic systems, energy systems, etc.) and the complexity triggered by us (natural catastrophes, technical catastrophes), greatly increases.

The value-creating contributions of complexity in these examples are unfortunately rarely noticed, and only the negative characteristics: generation of suspense and uncertainty, as well as of at-the-edge transition to chaotic forms.

However, complexity drives our progress. Without increase in complexity, there is no development. Yes, you can always go one step further. We owe our existence to increased complexity. Elementary particles combine under certain conditions to form atoms, atoms combine to form molecules, molecules into inanimate and living systems. From living systems conscious life eventually arises, and therefore humankind, as the most recent stage of development. If we think of this series of development from the beginning and into the future, development is always associated with complexity and a corresponding selection of system states. Selection is made possible by self-referentiality as described above.

As far as we know, the human brain is the system, with the highest level of complexity and far-reaching self-referentiality. Edelman and Tononi [48] outline a theory of consciousness which, as a fundamental idea, involves complexity and the integration of information on self-referentiality. Using Bateson's central theme of "the difference that makes a difference", they define a scientific concept for consciousness based on these ideas. What is referred to as "neural complexity", measures the ability to form integrated information via a mathematical formula. The "neuronal complexity" of the brain is low in childhood, greatest in adulthood and then decreases again in old age. Many neuronal cross-links are formed when we are toddlers, and then broken down again in favor of the formation of cross-linked neuron groups as we grow up. Groups of neurons form, within which a higher network of connectivity exists than in relation to the environment. The whole brain shows an additional coupling of "reentrant" networks that ensure "self-referentiality at the neuron level". This so-defined "consciousness" does not show a clear transition from the inanimate to the animate to conscious nature: consciousness already begins here with nonliving nature in the form of self-referentiality and by the selection of "right" states. This is a direct expression of our premise of not making a qualitative difference in fundamental models and theories of animate, animate, and social nature. This leads directly to the statement that although complexity increases, fundamental models and theories, which attempt to describe complexity, do not necessarily have to be different. As in the axiom, which was postulated at the outset, we find that fundamental scientific concepts are always used to describe conscious nature.

This finding is directly linked to the term "emergence" [19]. Emergence expresses the fact that the combination of elements creates something qualitatively new, which is not contained in the properties of the elements of a system, or in other words "the whole is more than the sum of the parts". Owing to emergence, nature takes leaps towards new qualities, from the inanimate to the living and then to conscious nature. The quality which emerges through the interactions of humans, the one that Luhmann calls the social system of communication, is a new stage of emergence.

It is important here that the connections between the elements at each stage of emergence are not arbitrary, but, according to Bateson, follow the axiom "the difference that makes a difference".

The interaction of the elements and the associated self-referentiality have another important effect. It is actually not meaningful to speak of elements, since elements as "things in themselves" or "beings in themselves" do not exist. This statement certainly has a constructivist component, but it is mainly due to the fact that the interaction of elements changes the elements. However, we will continue to speak of system elements, knowing that interaction generates a context for each element that brings out context-specific properties of system elements. We specifically talk of context-specific properties, because "properties per se" do not exist. Properties are only predispositions, which appear differently depending on context. Grawe's motivational schemes are a kind of predispositions (see Appendix "Fundamentals Consistency Theory").

Atoms contribute to a molecule's characteristics, but in molecules, new states are emerging, which produce new qualities in the molecule, and several molecules, as demonstrated in the example of water, produce further new qualities (see also the section "Regulation of Complexity Through Organizational Setting, Control and Order Parameters": The setting parameters of the environment have an influence on a developing macro-state and thus on the context-specific properties of a system). – We could wonder whether particle-wave dualism, for example, is the effect of an impermissible "property per se", which nature does not know as such. Psychology and social science have always opposed this unacceptable simplification: personality traits only represent a disposition, sometimes only a preference, and vary according to context. In the knowledge of this "unacceptable simplification", we nevertheless carry out this simplification, because it is very helpful for a first estimation. Social techniques, which are models, contain these simplifications and are thus not entirely free from reductionism. We believe that the alternative of not modeling such techniques at all, will lead to a form of insight exclusion.

In literature, the term "complexity" is often referred to in connection with the term "entropy", which is then (mistakenly) associated with disorder. Let us approach the connection of these concepts by explaining Fig. 2.7.

Let us begin by looking at a system in which the system elements have no interaction among themselves (in Fig. 2.7 middle section).

An example is an ideal gas of atoms in a container or a group of people in a room. In these examples, "the whole is the sum of its parts". Each system element shows its own behavior and it takes the micro-states it can occupy, due to the setting conditions. Since there is no networking, there is no complexity. A macro-state is completely determined by the system-element-specific micro-states. If there are many system elements and correspondingly many micro-states, a macro-state is formed by a set of the most probable micro-states. With an ideal gas in a container, this corresponds to a state represented by mean temperature and pressure. With regard to a group of people, especially since their number is never even approximately equal to that of gas particles, a corresponding statement is often difficult to achieve. If this is taken into account, we can observe, in a group of individuals who are not subjected to stress, more or less probable behavior of these individuals: For example, a large group of these individuals are sitting in a waiting room and not doing anything, another large group are reading. So here, too, we can speak, with good will, of a macro-state. Let us assume for a moment that the group's macro-state is well represented by the statement "99% of all individuals sit and wait in the

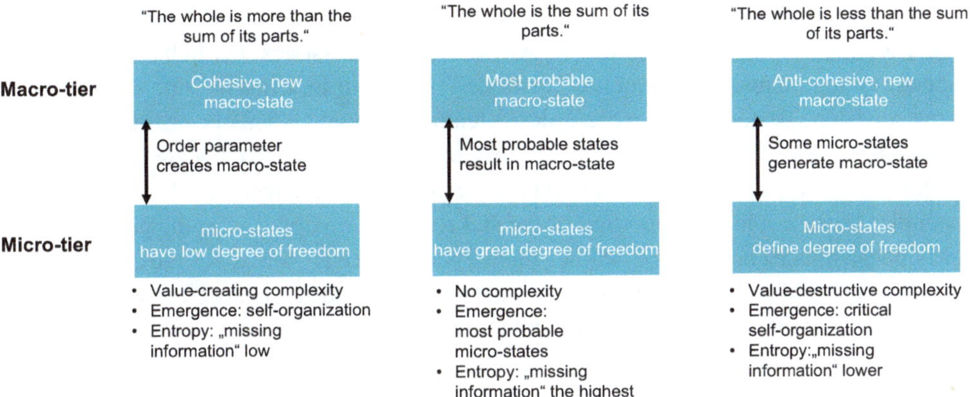

Fig. 2.7 Complexity and entropy

waiting room". In more detail, however, on a micro-tier there may be many different micro-states: A woman stretches her left leg at an angle of 30° and her right arm hangs down. A man bends forward, holding his head between his hands, etc. So we can see here that many micro-states can represent a macro-state. On the other hand, to be able to reason from a macro-state to a specific micro-state, much more information is required, information that is missing if we only know the macro-state. The famous law of entropy measures the quantity of micro-states, the micro-variety, and thus is also a measure for the "missing information" [49, 50]. Since the occupation of micro-states by system elements is also associated with energy expenditure, entropy is also a measure of energy that is "lost" by a system through the mere realization of micro-states. Very often, entropy is equated with disorder, but is connected with disorder, only to the extent that disorder is identified with a large number of (equally probable) states.

Complexity occurs when the system elements begin to interact with each other (left section of Fig. 2.7).

This interaction first costs energy and, under certain conditions, energy is released by the interaction and a cohesive new macro-state is achieved. At the same time, the number of micro-states is reduced, so entropy decreases, while complexity grows at first. Complexity grows as long as there are not too many connections that bring the system into a stable, static macro-state.

In a complex (social) system under certain conditions, a new dynamic macro-state may develop by self-referentiality. In this case, the possible micro-states, in the form of behavior, are restricted in a very specific manner and aligned to a new dynamic order. This process is called self-organization and the parameters that make this macro-state possible are setting, control and order parameters. Self-organization, in social systems, also means reduction of micro-states in favor of a dynamic macro-state. We assign a "value" to this macro-state, because it makes a new complexity level possible, on which further developments can take place. Therefore, in this case, we speak of value-creating complexity and say, "the whole is more than the sum of its parts".

Value-destroying complexity occurs when "the whole is less than the sum of its parts" (right section of Fig. 2.7). This occurs when, some micro-states, i.e. behaviors within a (social) system, cause the system to "explode" or remain in a macro-state that does not show cohesion. To give just two examples: a team "explodes" when one or more individuals fail to act in an appreciative way or are forging alliances outside the team to torpedo the team or project goals. Macro-states then emerge, which are recognizable as such, for example, because a large proportion of the stakeholders are opposed to the project. Other anti-cohesive states occur, for example, when a person actively torpedoes the co-operation, or only some of the team members participate actively, or the team speciate several (rival) groups, or the team cannot arrive at a collective agreement, because new ideas are constantly being created.

The formation of these three types of macro-states is significantly influenced by the nature of interventions in and around the system. Social interventions are stimuli or irritations by team members themselves or can be caused by external persons or organizations. The goal in project management should be to lead a social system from the macro-state "the whole is the sum of its parts", preferably without remaining long in the macro-state "the whole is less than the sum of its parts", into the state "the whole is more than the sum of its parts". In the section "Regulation of Complexity Through Targeted Interconnectedness and Self-Organization", we are therefore very much concerned with a set of interventions (eight principles), which allows the formation of this last highly emergent macro-state.

Literature

1. Simon FB (2009) Einführung in Systemtheorie und Konstruktivismus. Carl-Auer Verlag, Heidelberg
2. Beinhocker ED (2006) The origin of wealth. McKinsey & Company, Boston
3. Steeger O (2010) Mit Sozialtechniken "Prozessverlusten" entgegenwirken: Professor Lutz von Rosenstiel im Gespräch. projektManagement aktuell 5:3–9
4. Vollmer L (2016) Zurück an die Arbeit, Linde Verlag, Kindle Version
5. Luhmann N (1987) Soziale Systeme, 1st edn. Suhrkamp Verlag, Frankfurt am Main
6. Simon FB (2013) Gemeinsam sind wir blöd? Die Intelligenz von Unternehmen, Managern und Märkten. Carl-Auer Verlag, Heidelberg
7. Küpers W (2015) Die Kunst praktischer Weisheit in Führung und Organisation. In: Fröse MW, Kaudeler-Baum S, Dievernich FEP (eds) Emotion und Intuition in Führung und Organisation. Springer Gabler, Wiesbaden, Kindle Version
8. Singer T, Ricard M (2015) Mitgefühl in der Wirtschaft: Ein bahnbrechender Forschungsbericht. Albert Knaus Verlag, kindle Version
9. Wikipedia (2015) PDCA Zyklus. https://de.wikipedia.org/wiki/Demingkreis. Accessed 18 Oct 2015
10. Köhler J, Oswald A (2009) Die Collective Mind Methode. Springer, Heidelberg
11. Oswald A, Köhler J (2012) Mit dem Projektnavigator zum Projekterfolg. In: Lang M, Kammerer S, Amberg M (eds) Perfektes IT-Projektmanagement. Symposion Verlag, Dusseldorf, pp 167–192
12. Oswald A, Mourgue d'Algue H, Merk G (2014) Fundamental requirements for a framework of adapted theories for project work. In: Rietiker S, Wagner R (eds) Theory meets practice in projects. GPM Deutsche Gesellschaft für Projektmanagement e.V., Nuremberg, pp 52–65

13. Rietiker S, Scheurer S, Wald A (2014) A framework for theory-based problem-solving. In: Rietiker S, Wagner R (eds) Theory meets practice in projects. GPM Deutsche Gesellschaft für Projektmanagement e.V., Nuremberg, pp 80–92
14. Winter M, Szczepanek T (2009) Images of projects. Gower Publishing
15. Hanisch B, Wald A (2013) Projekte als geeignet Organisationsform zur Handhabung von Komplexität: Ergebnisse einer Studie zu Beratungsprojekten. In: Wald A, Mayer T-L, Wagner R, Schneider C (eds) Komplexität. Dynamik. Unsicherheit. Advanced Project Management, vol 3. GPM Buchreihe Forschung, Buch Nr. F07. GPM Deutsche Gesellschaft für Projektmanagement e.V., Nuremberg, pp 127–142
16. GPM Deutsche Gesellschaft für Projektmanagement e.V. Tabelle zur Bestimmung der Komplexität eines Projektes, http://www.gpm-ipma.de/qualifizierung_zertifizierung/ipma_4_l_c_zertifikate_fuer_projektmanager/level_b.html, Unterlagen zur Erstzertifizierung IPMA Level B: Zertifizierter Senior Projektmanager (GPM)
17. Santa Fe Institute (2014) Complexity explorer. Online Kurs Introduction to Complexity. http://www.complexityexplorer.org/courses. Accessed 20 Sept 2014
18. Dittes F-M (2012) Komplexität: Warum die Bahn nie pünktlich ist. Springer, Berlin/Heidelberg
19. Greve J, Schnabel A (eds) (2011) Emergenz: Zur Analyse und Erklärung komplexer Strukturen. suhrkamp taschenbuch wissenschaft, Berlin
20. Mainzer K (2008) Komplexität: UTB Profile. Wilhelm Fink Verlag, Paderborn
21. Stoica-Klüver C, Klüver J, Schmidt J (2009) Modellierung komplexer Prozesse durch naturanaloge Prozesse. Vieweg+Teubner Verlag, Kindle Version
22. Ashby WR (1957) An introduction to cybernetics. Chapman & Hall, London
23. Grawe K (2004) Neuropsychotherapie. Hogrefe, Göttingen
24. Oswald A, Köhler J (2010) Wechselwirkende Organisationen, Teil 1. projektManagement aktuell 5:14–19
25. Oswald A, Köhler J (2011) Wechselwirkende Organisationen, Teil 2. projektManagement aktuell 1:36–41
26. Cavanagh M (2011) Complex or merely complicated? https://www.youtube.com/watch?v=MBgNrVfpx38. Accessed 14 Feb 2016
27. Gleick J (2011) Chaos: making a new science. Open Road Media, Kindle Version
28. Dörner D (2011) Die Logik des Misslingens: Strategisches Denken in komplexen Situationen. rororo Verlag, Kindle Version
29. Appelo J (2011) Management 3.0: leading agile developers, developing agile leaders. Pearson Eduction, Boston
30. Haberstroh M (2013) Angepasstes Projektmanagement bei Unsicherheit und Dynamik. In: Wald A, Mayer T-L, Wagner R, Schneider C (eds) Komplexität. Dynamik. Unsicherheit. Advanced Project Management, vol 3. GPM Buchreihe Forschung, Buch Nr. F07. GPM Deutsche Gesellschaft für Projektmanagement e.V., Nuremberg, pp 174–195
31. Shenhar AJ, Dvir D (2007) Reinventing Project Management: The Diamond Approach to Successful Growth & Innovation. Mcgraw-Hill Professional, Boston
32. Elsberg M (2012) BLACKOUT – Morgen ist es zu spät. Blanvalet Verlag, Kindle Version
33. Bak P (2014) How nature works: the science of self-organized criticality. Copernicus, Kindle Version
34. Haken H, Schiepek G (2010) Synergetik in der Psychologie: Selbstorganisation verstehen und gestalten. Hogrefe-Verlag, Göttingen
35. Parongama S, Chakrabarti BK (2013) Sociophysics: an introduction. OUP Oxford, Kindle Version
36. Weidlich W (2012) Sociodynamics: a systematic approach to mathematical modelling in the social sciences. Dover Publications, Kindle Version

37. Fromm J (2004) The Emergence of complexity. Universität Kassel. http://www.uni-kassel.de/upress/online/frei/978-3-89958-069-3.volltext.frei.pdf. Accessed 14 Feb 2016
38. Oswald A, Köhler J, Schmitt R (2015) Wertschaffende Komplexität und Selbstorganisation. projektManagement aktuell 4:45–51
39. Snowden DJ, Boone ME (2007) A leader's framework for decision making. Harv Bus Rev (November): 1-8
40. Gigerenzer G (2014) Risiko: Wie man die richtigen Entscheidungen trifft. C. Bertelsmann Verlag, Kindle Version
41. Wikipedia (2015) Risiko, Ungewissheit und Unsicherheit in der Entscheidungstheorie. https://de.wikipedia.org/wiki/Entscheidungstheorie. Accessed 18 Oct 2015
42. Wikipedia (2015) Known knowns, unknown knowns, known unknowns, unknown unknowns. https://de.wikipedia.org/wiki/There_are_known_knowns. Accessed 18 Oct 2015
43. Taleb NN (2012) Antifragile. Things that gain from disorder. Random House, New York. Kindle Version
44. Bardmann TM, Lamprecht A (1999) Systemtheorie verstehen, eine multimediale Einführung in systemisches Denken. Westdeutscher Verlag GmbH, Wiesbaden
45. Von Bertalanffy L (2006) General system theory, Revised edn. George Brazillier, New York
46. Weinberg GM (2001) An introduction to general systems thinking. Silver Anniversary Edition. Dorset House Publishing, New York
47. Bateson G (2014) Geist und Natur, eine notwendige Einheit. suhrkamp taschenbuch wissenschaft, Frankfurt am Main
48. Edelman GM, Tononi G (2013) Consciousness: how matter becomes imagination. Penguin Press Science, Kindle Version
49. Ben-Naim A (2008) Entropy Demystified. Verlag WSPC, Kindle Version
50. Hägele PC (2004/2005) Was hat Entropie mit Information zu tun? Universität Ulm. http://www.uni-ulm.de/~phaegele/Vorlesung/Grundlagen_II/_information.pdf. Accessed 14 Feb 2016

3 Options of Complexity Regulation

The best in this kind are but shadows; and the worst are no worse, if imagination amend them.

William Shakespeare, A Midsummer Night's Dream

After having learned the essential properties of complex systems in previous chapters, we will now approach the regulation of complexity in projects in this chapter. We will do this by addressing the fundamental possibilities of regulation. We need these as a prerequisite to classify, understand and apply the theories, models and examples listed in the following chapters.

What are the main ways at our disposal for using complexity to create value? We do not use the term complexity management in this context, because it is misleading, as complex systems cannot be managed, although a particular form of leadership is possible. According to Ashby [1], when we speak of regulation we mean any modeling of the complexity of a system to influence its complex behavior. Influence here means exerting deliberate stimuli or interventions on the complex system "project" in order to stimulate its development in the sense of the goal.

A. Oswald et al., *Project Management at the Edge of Chaos*,
https://doi.org/10.1007/978-3-662-48261-2_3

Complexity: Reduce or Better Regulate?

Ehrlich and Priesberg are on their way to a lecture on complexity management. One of their colleagues declares: "Finally we will be able to manage complexity away. Another problem solved."

Ehrlich mutters angrily to himself on the way to the lecture: "We broke through the sound barrier, surely we can break through the speed of light barrier, it is all a question of management."

Priesberg interrupts him: "I really do not understand you. Complexity can be reduced by targeted steps and therefore mastered at the end of the day."

Ehrlich groans and clasps his hand over his forehead: "Have you been playing Buzzword Bingo lately? If it is as simple as you say, then the concept of 'complexity' would only exist in museums – everything would then be reduced. So, you have to get used to one thing: As soon as you enter the grounds of complexity, the familiar pattern of cause and effect is no longer valid. In the case of complex contexts, you are stepping away from everything that is going on and you are looking for control parameters with which certain order parameters can be adjusted. And you will have to expand your intuition considerably – the discussion with the speaker will be fun if he wants to sell us the reduction of complexity," he concludes, rushing to grab the few vacant front seats in the auditorium – from where he can contradict the speaker directly …

We see the following possibilities of regulation:

Regulation through shielding in space and time by controlling the flow of social complexity in and out of a temporary organization "project". In many cases this corresponds to a reduction in complexity within a project.

Regulation through the formation of models and intuition is regulation in the narrower sense according to Ashby and Conant by the formation of models corresponding to complex systems [1, 2, 3]. Using a variety of examples, Dörner was one of the first to show the practical importance of models for understanding complex systems [4]. This modeling may refer to complex technical or social systems, but may also explicitly express the interaction of social and technical systems. Later on, we will show that intuition is a key tool for regulating complex situations, and that experiences emanate from the application of models, which serve to develop model-based intuition.

Regulation through targeted interconnectedness and social self-organization, is a form that clearly goes beyond the possibilities of the regulation of a single individual. A team forms a collective mental model (the Collective Mind) to adequately model the complexity of a project's task through its team variety. This gives the team a unique position in the regulation of complexity. The formation of the Collective Mind is a cohesive emergent macro-state caused by self-organization within the team.

Regulation through organizational setting, control and order parameters is a form of regulation, which specifically uses the design of these parameters to regulate complexity in an organizational network. Individuals and organizations are represented through the Dilts pyramid model [5, 6] and parameters are assigned to the Dilts pyramid logical levels.

3.1 Regulation of Complexity Through Shielding in Space and Time

...
Adrian: The air breathes upon us here most sweetly.
Sebastian: As if it had lungs, and rotten ones.
Antonio: Or as 'twere perfum'd by a fen.
Gonzalo: Here is everything advantageous to life.
Antonio: True; save means to live.
Sebastian: Of that there's none, or little.
Gonzalo: How lush and lusty the grass looks! how green!
Antonio: The ground, indeed, is tawny.
...

William Shakespeare, The Tempest

The regulation of complexity through shielding in space and time has a long tradition in project management, although it has not really been described as such: Withdrawing a project team, e.g. in the context of a kick-off meeting, the aim is to achieve peace away from daily business and associated disturbances. The formation of a core team and an extended team also have a corresponding complexity-regulating function. The existence of too many team members with different personalities and interests can lead to "disordered" interaction. This also involves rules for the fluctuation of team members in workshops or meetings, as too frequent changes in team composition lead to a high level of complexity import.

Subtle forms of complexity regulation include settings in workshops. For example, an open circular chair arrangement, or a seating arrangement with tables set out into a U-shape are completely different complexity drivers. A circular chair arrangement will increase social complexity if some workshop participants feel uncomfortable. Conversely, seating arrangements with tables can help to reduce complexity, because this gives some participants a feeling of safety and order.

The shielding or deliberate toleration of complexity in space and time can fulfill various functions:

Through shielding in space, stable setting conditions are created, and rest, safety and focus set in. Conversely, the deliberate opening of spaces (including large-scale spaces, interaction spaces) leads to an admission of complexity, assuming that within the defined space a fruitful interaction arises.

If the shielding is open for a specific time period, this time can be used for information retrieval. Consultation with the extended team has this function for the core team. Targeted communication with other stakeholders, e.g. customers, also fulfills this function.

A deliberate toleration of complexity in a team, where some team members even talk of perceived chaos, can speed up team building. As we shall see later, a minimum of confusion and stress is actually necessary to trigger change and learning.

In Agile Management, shielding in space and time is a key principle. In the agile framework of action, Scrum, the product owner is (ideally) the only gate from the team to the customer. Complexity is channeled through the product owner. The reduction of work-in-progress (WIP) serves to "block out" complexity for team members. Even for an organization using the agile frameworks Scrum and Kanban or the Critical Chain method in project management the WIP limitation fulfills the same function [7]. Work-in-progress acts as a control parameter.

Figure 3.1 illustrates this fundamental idea of regulation by shielding: The boundaries of the temporary organization "project" can be partially open at the time t (n) (e.g. some of the team members have contact with other stakeholders) and further or more narrowly defined as required (e.g. the circle of team members is expanded). At a later point in time t (n + 1), it may be necessary to deliberately close the boundaries of the project in order to give the team the opportunity for self-organization. We will see later that stable "project boundaries" are a central prerequisite for the formation of macro structures by means of self-organization.

Within a team, it may also be necessary to design individual complexity processing of team members by means of a leadership oriented to team members' individual basic needs. Different personalities have different needs and abilities for processing complexity. A key task of leadership is to shield complexity for each team member individually. We will also return to this issue in the section entitled "Regulation of Complexity Through Targeted Interconnectedness and Self-Organization".

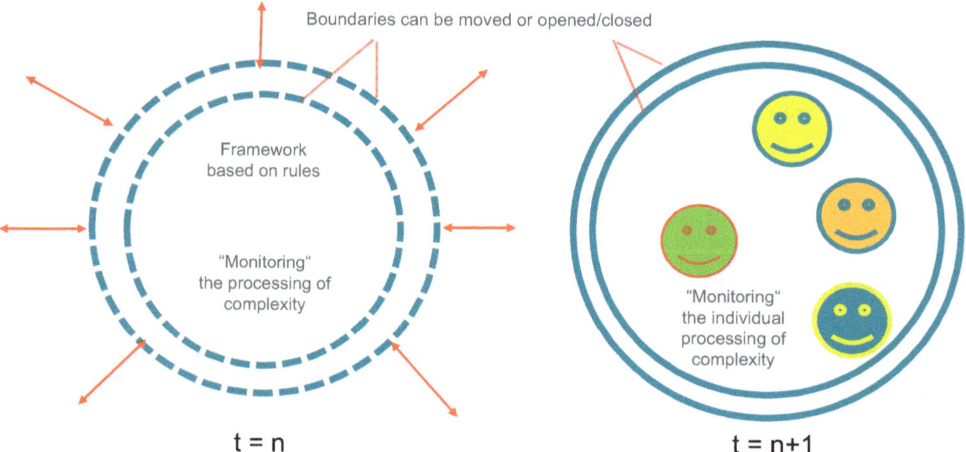

Fig. 3.1 The open system "project"

3.2 Regulation of Complexity Through the Formation of Models and Intuition

...Ideas always come up out of the water – where else would they come from? But they go away if we talk about them. So say nothing, but go on tapping.

Fred Vargas, A climate of Fear,
English by Siân Reynolds

The regulation of complexity and associated actions under uncertainty are based on the following fundamental insights:

In order to regulate complex systems, it is necessary to form models of these systems. These models must have a variety that is at least equal to that of the systems themselves. This insight can be traced back to Ashby [1] and Conan and Ashby [2]. It confronts us with a huge problem: Models need to be constructed that exceed the variety in our project world. This should be impossible. As Conan and Ashby demonstrate, our mind is a (imperfect) model of our environment and yet we are (more or less) able to shape our own lives. By counting the dimensions of the models used in this book, we can quickly clarify that the variety depicted by the integration of all models is enormous. Demand should not be to find the perfect model, whose variety is greater than the variety of the world itself. Instead, it is useful to define the right scale and then test suitable models and their varieties in practice: Examples of manageable models with sufficient variety are models of personality, culture of organizations, stakeholder management, project types and complexity classes. In finding the right scale, however, emergent macro-states also come to our aid. This is because Ashby's theorem does not make any explicit statement regarding the system level to which it may be applied. Since the macro-tier contains fewer states, we are in a much better position to model it appropriately and to regulate it through interventions. We will use this knowledge again and again in the following chapters.

Since neither unconscious or conscious models, which we only acquire through experience, and not using explicit models and theories, do not usually have the variety necessary for the regulation of a complex system, complex systems cause uncertainty in our actions. As we will see in the section "Interplay of Intuition and Rationality", the gathering of detailed information on the system is important to avoid mental distortions, but not really helpful for making a decision. Instead, it is useful to refer to explicit models that represent the macro-states of complex systems. In the simplest case, these models are rules of thumb (heuristics) that have been gained through experience in a particular context, and are provided through intuition in uncertain situations. This insight is based on the work of Kahneman and Tversky [8] and Gigerenzer [9, 10]. We go a step further here by stating that deliberately designed models of system behavior are "better" than simple rules of thumb (i.e., they do not produce simple distortions), because they involve much more variety. Often, however, we rely on rules of thumb, especially in stress situations, because their results reach our consciousness through intuition, but the processing itself is unconscious. Edelman and Tononi [11] describe this interplay of conscious and unconscious neural processes exemplarily using the example of learning a musical instrument: "Techniques" are regularly used and are successively transformed into automated and unconscious processes. In the same way, consciously constructed models should be regularly used and tested in practice to train intuition as unconscious competence with more variety. The basic idea, to use models consciously in order to improve learning under conditions of uncertainty, is contained in the book "Theory meets Practice" [12] and is the motivation to deal with a theory-framework for project work. Intuition is thus not developed by a theory or a model, but by a multitude of interlinked theories and models, which have also been adapted for project management. And that is precisely the central concern of this book.

The rules of the thumb, which underlie our intuition, are not excluded as a source for models, as long as their context dependencies are taken into account. To summarize, we can say: "The design of models is based on rationality, while that of operational applications is based on intuition." Techniques of improvisation also reflect this insight: Good improvisation lives from the balance of a more abstract theoretical understanding of system structures and practical experience, which has been acquired through a lot of practice in situational actions [13].

As we have seen above, complex systems show comparatively simple behavior when forming macro structures. In such cases, it is not necessary or useful to influence the behavior of the complex system by regulating it at the micro-tier, since this can be done at the macro-tier. The order parameters of the macro-tier act in a regulating manner on the micro-tier: for example, values, beliefs, and rules of thumb act as order parameters on human behavior. A person who considers it important to have order in their lives will create order in their behavior. In general, order parameters may be subject to an order parameter hierarchy. Stafford Beer's Viable System Model [14] and the NLP Dilts Pyramid [5, 6], are prominent examples of order parameter hierarchies. The target hierarchy of the Collective Mind model is a 3-step order parameter hierarchy for goal-oriented project management. Figure 3.2 shows the Dilts Pyramid:

Dealing with Complexity Means Transformation Work for All

Priesberg meets Ehrlich in the elevator. "I've got a new boss and I'm wondering why he did not want any details about my work area, but instead asked about the principles and foundations of my actions. So, he will be unable to judge my work. In addition, he has also worked out a new behavioral code with us".

Ehrlich ponders: "Let me guess: After the rules have been agreed and passed, resistance has become widespread during the first implementation." Priesberg nods in agreement and Ehrlich continues: "Your boss seems to be a person who leads in a value-oriented way and then these values finally determine what is worked out in detail. It is a very powerful principle and is at the very core of self-organization. However, your new boss has to provide the time and means of learning for all those involved, so that everyone is able to internalize his approach. Otherwise they will be unable to go along with it."

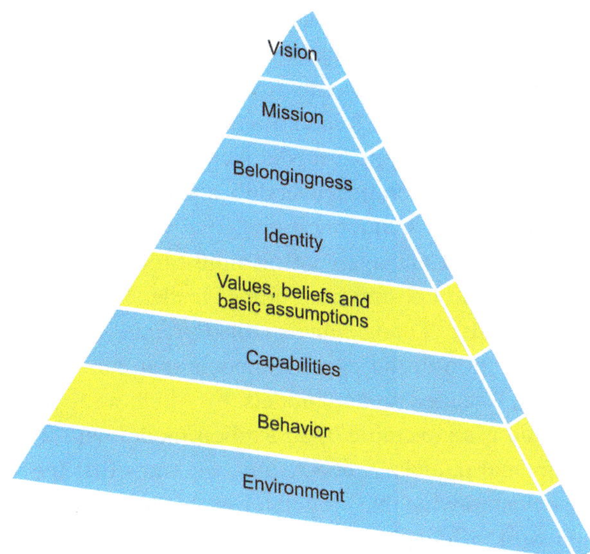

Fig. 3.2 Dilts Pyramid

The central statement of the Dilts Pyramid is that the respective upper level "regulates" the underlying levels: Any communication (behavior) takes place in a particular environment or context. Communication is shaped by communication capabilities. Values, beliefs and basic assumptions influence behavior through respective communication abilities. Communication blockades come up, for example, due to values and beliefs that have been meaningful and successful in certain past contexts, but are inappropriate for the current context. In order to achieve a sustainable change, in the case of communication blockades, it is necessary to change the order parameters "values, beliefs and basic assumptions". This applies to both personal and organizational changes. The geometrical figure "pyramid" demonstrates that the variety of possible characteristics increases towards the bottom of the pyramid. There are numerous contexts, with many behaviors (micro-states) that a human being can demonstrate. The number of capabilities is already much lower and only a few values, beliefs and basic assumptions control this behavior. The rules of thumb (heuristics) belong to this level of values, beliefs and basic assumptions and thus form order parameters for our behavior. Therefore, although a human being can show numerous behaviors, and a team or individuals in an organization even more so, regulation of this behavior makes sense through the order parameters of the higher levels of the Dilts Pyramid. The degree of abstraction[1] increases with the higher levels, but at the same time

[1] We understand abstraction as a form of bundling or class formation: different behaviors are assigned to a capability, the capabilities are themselves assigned to a value, a belief system or a basic assumption, etc. In this case the bundles or categories become more abstract with each level of bundling or class formation. The value of "trust" in this sense is much more abstract than one of the assigned abilities such as "approaching other people with an open mind" and this capability is, in turn, more abstract than concrete behavior: "While he is approaching the other person, he smiles and waves happily".

the chance to regulate a complex system with relatively simple means increases. Einstein summarized this ingeniously: "We cannot solve our problems with the same level of thinking that created them." This form of regulation through higher levels of abstraction also allows for maximum freedom on the micro-tier, i.e. in human action. The Dilts Pyramid thus represents a simple model for depicting different leadership styles – from visionary leadership to autocratic leadership, and their consequences (see the section "Fundamentals of Leadership"). Guideline-oriented governance concepts in society, in organizations and also in project management or multi-project management, take advantage of this, via a higher levels of abstraction form of leadership. We shall return to this in more detail later (see appendix "Fundamentals Dilts Pyramid").

In addition to the explicit models that can be found in the literature, there are also models, so-called mental models, by which we represent the world in our brain. The Dilts Pyramid is an example of an explicit model with which we elucidate the mechanisms of our mental models. Since Ashby, we know that the variety of the regulator is crucial to the extent to which this regulator is able to regulate a complex system. In the transformation work of people, as well as in the transformation management of organizations, we speak of whether resources are present to cope with the situation. As we all know, a vision, in addition to values and beliefs, can be very helpful in coping with a situation or achieving a goal. However, many of us may have experienced that a vision, a value proposition or a belief system can also have a blocking effect to a great extent, which binds our resources or restricts us in our variety. Psychological or social order parameters develop a great force, but not always for our own benefit. Think of the "belief wars" of different technologies in the IT environment. How many projects may have already failed for that reason!

Different people, as well as organizations, have different (mental) flexibilities, and therefore different abilities to create mental variety. This involves a subjective view that different people feel the same system to be more or less complex. In this way, however, a system is objectively no more or less complex, it is only our ability to model this complexity with our variety, which makes the system seem more or less complex.

Personal transformation work or transformation work in organizations therefore, is to increase respective variety. For example, personality models are tried and tested means of stimulating and sharpening self-reflection, which makes our own behavior more easily recognizable by patterns and also allows potential behavior change. Variety, which has been cultivated in this way, in combination with experience, should be at a considerably higher level than that attained individually through the experience created rule of thumb.

The well-known French crime author Fred Vargas created a detective, Jean-Baptiste Adamsberg, who solves tricky murder cases based on his intuition and with the help of his rational colleague Adrien Danglard.

The head of a project management office complains of the ineffectiveness of more or less complicated PM tools, instead of advising their project leaders to trust their intuition in complex projects.

The Nobel laureate Daniel Kahneman, who has scientifically studied the disadvantages and advantages and patterns of intuition, advises all those who, in answer to their question of how the person they are speaking to came to their opinion, receives the answer "intuition", to distrust this opinion.

The MPI Professor Gerd Gigerenzer, is so convinced by the underestimated importance of intuition that he advises us to trust our own intuition more and to use intuition to solve problems.

On the basis of these examples and our own experiences, we use the term intuition in four different contexts:

- Intuition in transcendent processes
- Intuition in automated processes
- Intuition in decision processes
- Intuition in creative processes

Intuition in transcendent processes points beyond us and is the access to ourselves and to our connection with the universe. In this context, intuition is the means to find the meaning of our life. In the Dilts Pyramid, this form of intuition begins at the top three neurological levels and points beyond these.

Intuition in automated mental processes means the execution of unconsciously carried out activities, such as driving, reading, talking, estimating the effort of tasks in projects, and many more. In all these examples, we have built up unconscious competence after a certain period of practice. We do not have to think about what and how. As soon as we think, we are slower and make more mistakes. We can even deliberately evoke this, for example by asking a driver, in the middle of a difficult maneuver, to think consciously, and the driver will then be unable to master the situation as well as otherwise.

Intuition in the decision-making processes or in the context of the management of systems and organizations has attracted much attention in recent years. As Gigerenzer points out, in a business, intuition grows in importance as a mean of decision-making with the hierarchy level [9]. Daniel Kahneman received a Nobel Prize for Economics in 2002, because of his investigations, together with Amos Tversky, into the effects of intuition on the decision-making processes [8]. He arrived at the conclusion that intuition is a powerful ability, but does not always entail positive consequences. Gerd Gigerenzer partly agrees with this view, he emphasizes that intuition is often underestimated in its effectiveness. In many cases, he prefers the simplicity of intuition in the form of rules of thumb over the intricacy and complexity of management systems [9, 10].

Kahneman concurs with his colleague and describes intuition as follows [8]:

The situation has provided a cue; this cue has given the expert access to information stored in memory, and the information provides the answer. Intuition is nothing more and nothing less than recognition.

Gigerenzer uses the following definition [9]: "Any rule of thumb known to me can be used consciously or unconsciously. If it is used unconsciously, we speak of intuitive judgments. An intuition or gut feeling is judgment that emerges abruptly in consciousness, whose deeper reasons we are not quite aware of, but it is strong enough to act on."[2]

Furthermore he says "… It (the gut feeling[3]) is a form of unconscious intelligence".

Gigerenzer mentions the following examples of rules of thumb:

- Less is more.
- Listen first, then speak.
- If a person is not honest and trustworthy, the rest does not matter.
- Innovation is the driving force behind success.
- You cannot bet on security and win. Even an analysis will not reduce suspense.
- If a customer has not made a purchase for 9 months or more, they are considered inactive, otherwise active.
- Good leadership consists of a toolbox with rules of thumb and the intuitive ability to quickly identify which rules are appropriate to which context.

It is easy to see that rules of thumb are nothing more than basic assumptions, which may also manifest as beliefs. From our consulting and professional practice, we list a few basic assumptions which we consider to be beliefs, and which subsequently may not be valid everywhere:

- Customer satisfaction over employee satisfaction.
- Better to make a wrong decision than none at all.
- You get what you are looking for.
- Innovations are created by customer requests.
- A good project leader is one who masters their methodical tools.
- Anyone can become an excellent project leader.
- Freedom is the most important quality.
- The customer gets what they need, not what they want.

Since intuition plays a prominent role in decision-making processes in project management, we will deal specifically with this form of intuition in the following section "Interplay of Intuition and Rationality".

The fourth form of intuition "intuition in creative processes" makes it possible to search for connections in a murder case for example, to find an innovative project solution, or to create theories and experiments in research and development.

Albert Einstein is considered to be an absolute master of creative intuition. Since creative intuition plays an important role in innovative projects, we look at the related mental

[2] Translation by the authors.

[3] Annotation by the authors.

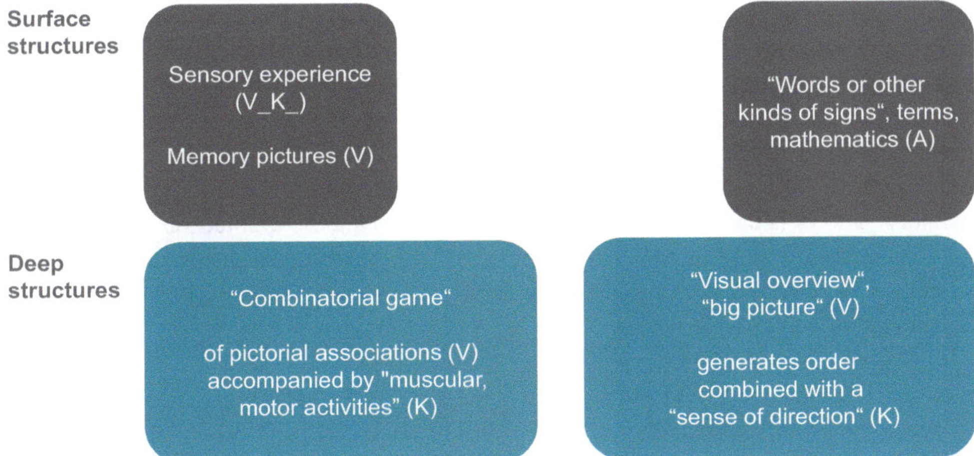

Fig. 3.3 Einstein's mental strategy

processes somewhat more closely. In his very accessible book on Einstein [15], Dilts analyzes his mental strategy[4] and gives Einstein a voice:

> "On the stage of our mental experience, **sensory experiences** and memory pictures of such ideas and feelings appear in colorful succession. In contrast to psychology, physics deals directly with the sensory experiences and the "comprehension" of the relationship between them. But the concept of the "real outside world" of everyday thought is also based exclusively on the sensory impressions." … " The **words of the language**, as they are written or spoken, do not seem to play any role in my way of thinking. The mental structures which seem to me to serve as elements of thought are definite signs and more or less clear pictorial representations which can be 'deliberately' reproduced and combined. Of course there is a certain relationship between those elements and relevant logical terms. It is also clear that the emotional basis of this rather vague game with the mentioned elements is the desire to ultimately arrive at a logically coherent system of terms. From a psychological point of view, however, this **combinatorial game** appears to be the key feature of productive thinking." … "During all these years I had a **sense of direction**, the feeling of going straight to something specific. It is, of course, very difficult to express this feeling in words, but it was decidedly the case, and clearly distinguishable from the nature of the later contemplation of the rational kind of solution. Of course there is always something logical behind such a direction; But I've got it in some **kind of overview**, virtually visible in front of my eyes."[5]

Based on this quotation and the Dilts' analysis, the model of Einstein's mental strategy is sketched in Fig. 3.3 results.

The first step in his strategy is the intentional generation of sensory experiences through the senses V (visual) and K (kinesthetically), combined with memory images. He then

[4] In line with NLP [6], we are talking about a mental strategy to make it clear that it can be related to different intuitive contents and can also be used by others as a model for the training of creative intuition.

[5] Translation by the authors.

speaks of pictorial associations and muscular, motor activities which unconsciously trigger a combinatorial game with many psychic and physical parts. These parts produce a big picture, which in turn creates order, combined with a sense of direction. And only when this big picture exists, will words, terms and signs, etc., from mathematics, be used to mold the picture into formal structures. The big picture is the order parameter, which gives order to the combinatorial game. The intentional induction of sensory experiences and memory pictures is the control parameter, which allows the use of intuition.

The intuitive strategy of problem solving used by Einstein corresponds precisely to the strategy of creative problem solving recognized by Beck [16], according to neuroscientific insights:

- Conceiving and specifying the problem in the prefrontal cortex (with Einstein, this happens with visual and kinesthetic sensory experiences).
- If the (rational) activity in the cortex is insufficient to solve it, the problem is divided. Depending on the type of processing (visual, auditory, kinesthetic, olfactory, gustatory) different brain areas play a part. Einstein calls this parallel unconscious processing the combinatorial game.
- After a certain time[6] of separated processing, the distributed centers report their results as an activity pattern to the prefrontal cortex. There is a search for order in the activity patterns. If an order can be established by means of an ordering parameter (an integrated information), a new conscious thought exists. Note the agreement with Einstein's description of a "big picture" and sense of direction, and also Edelman and Tononi's theory of consciousness.
- Subsequently, the flash of inspiration obtained in this way, is intentionally poured into literature, paintings, gourmet menus or into music or mathematics. In most of these cases, this 4-step process is likely to be repeated many times to provide excellent results.

If we compare this concept with intuition in the decision-making process, the process is very similar, albeit simpler. Sensory images and memory images provide mental clues, which are usually unconsciously used to recall rules of thumb or basic assumptions made on the basis of prior experience. In addition, this leads to a decision: A rule of thumb, a heuristic technique, is synthesized from different, but sufficiently similar situations (this corresponds to sensory experiences along with the combinatorial game). Very often, however, distortion takes place, i.e. the rule of thumb does not fit as a solution, as a "big picture", to the situations in which it is applied. If this non-appropriate rule of thumb is applied, bad decisions are made. Kahneman illustrates the associated perils. However, if the rule of thumb is appropriate for the situation and does not contain any distortions, there is a tremendous advantage, because simple rules can be used to regulate complex situations. And it is precisely this that is "the faith" referred to by

[6] Some also speak of an incubation period: Here, wrong tracks may be taken or mistakes made, as these increase the incubation time. Within the framework of the theory of self-organization, we speak of fluctuations that are necessary to scan the space of possibilities.

Gigerenzer, which he has underlines with many examples. One of his favorite rules of thumb is "less is more" and he lists many examples in which complex management systems have caused a great deal of damage. For example, he is convinced that simple rules of thumb bring more beneficial effects to global financial management than value-destroying management systems.

In the language of complexity, this means that we unconsciously extract patterns from experiences and create order parameters on their behalf. These order parameters appear in the form of basic assumptions, beliefs, or rules of thumb.

We will see later that this approach can also be applied in teams: In visual and, possibly, kinesthetically oriented workshops, a "big picture" emerges in the team after an extensive combinatorial game of formulating goal approaches. This "big picture" is concretized on different levels and integrated iteratively with the elaborated details. The "big picture" acts as the topmost order parameter for the underlying sub-aspects.

In summary, we can say that intuition is cultivated when experience patterns have formed in our minds and through memories, which can be recalled, depending on the situation or through creative association. Therefore, it is important to recognize any distortions early on. It is here that rationality can help.

However different these various kinds of intuition may be, in all cases, our body seems to play an important role. When we speak of intuition, we almost always speak of good or bad feelings, gut feelings or other physical reactions. Damasio speaks of "somatic markers" [17], which accompany our cognitive processes and suggests preference by means of physical signals.

We also observe that people have different levels of ability of intuition, so that there are different personalities with different manifestations of and preferences for using intuition. When we consider Einstein's example, we find that intuition creates excellence when intuition and rationality interplay with brilliance.

3.2.1 Interplay of Intuition and Rationality

If there was one thing about Adamsberg that Danglard disapproved of more than anything else, it was his habit of considering his hunches as solid fact. Adamsberg regulary replied that hunches were facts, material elements as valuable as a lab test.

Fred Vargas, The Ghost Riders of Ordebec,
English by Siân Reynolds

In Kahneman and Tversky's model [8], the interplay of intuition and rationality is represented as fast and slow thinking of two mental systems: System 1 (intuition) and System 2 (rationality). We have elaborated upon, and transferred the fundamental ideas of Kahneman and Tversky [18] to projects, which we have adopted for this section.

Intuition and Rationality

Priesberg cannot get the lecture on complexity management out of his mind. Apparently, it was not as bad as Ehrlich had feared. "Do you really think that our thinking can be separated into two areas, fast and slow thinking, as mentioned in the lecture?", he asks. Ehrlich grins: "I'll give you a task: A bat and a ball cost €1.10. The bat costs one Euro more than the ball. How much does the ball cost?" Priesberg immediately replies: "10 Euro cents." "False," Ehrlich beams widely, "you should have pondered a little longer. The solution is 5 Euro cents. This example beautifully illustrates the way fast and slow thinking works. So you see, perhaps there is more to it than meets the eye," he concludes.

As Ehrlich and Priesberg have already found out in the above example of "intuition and rationality", the answer that most people come up with, is 10 Euro cents. This answer comes to our mind quickly, but is wrong. If you take a piece of paper and solve the task, you will discover that the correct answer is actually 5 Euro cents.[7]

Those who stated 10 Euro cents, were misled by their intuition, i.e. System 1. Those who provided the correct result have probably deprived themselves of the influence of System 1, by subjecting quick results to a slow check and solving the equations listed in the footnote. System 1 sent a signal to System 2: For some, this signal is simply "rubber-stamped", for others, it undergoes active control and is corrected according to the findings.

Another example from Kahneman's book aptly illustrates the influence of intuition. When, under test conditions, individuals are provided with a screen saver with floating dollar bills, they start to show a statistically significantly higher degree of independence, and exhibit more selfish behavior: This effect is an example of the so-called priming effect and is evoked by System 1. Experimental findings suggest that mental canalization (priming) strengthens individualism through the "idea of money" and thus leads to a corresponding bias of intuitive decisions (for example in the direction of maximizing one's own income). In a critical project situation, in which the project leader places special emphasis on the financial consequences, team spirit may not be supported, but individualism is unwittingly strengthened. This can lead to an increase in social complexity as team members interact in a way that brings the project into crisis.

Intuition is generally very susceptible to any form of mental distortion and, for this very reason alone, is only conditionally suited to making good decisions under conditions of uncertainty. At the same time, the strengths that result from intuition are also weakness: As intuition involves refraining from careful examination of data and facts, it is therefore, a fast reaction. Since, in the case of uncertainty, there is either no data available, or it is difficult to access, intuition provides explanations based on easily accessible information. These can be readily accessible mental models, caused by priming, or based on experiences in the form of rules of thumb. These patterns, generalized and abstracted from past experiences,

[7] By solving the two equations, you get the result of 5 Euro cents:

Price bat + Price ball = 1,10 €

Price bat = 1 € + Price ball

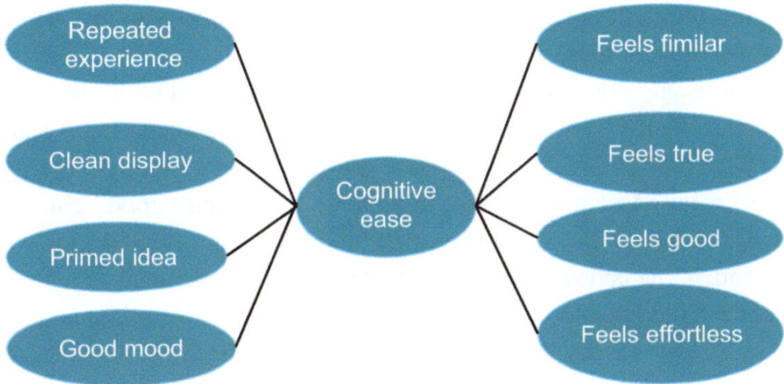

Fig. 3.4 Cognitive ease and intuition

can only help master an uncertain situation if they are not too alien to the new, trouble causing, structures: An experienced project leader will react appropriately and "in the right way" when a new project situation is similar in its characteristics to previously experienced project situations.

We all tend to yield to cognitive ease. Cognitive ease occurs when a repeated experience is present, a clear display or already primed ideas are offered or a good mood underlays the situation. In these cases, the situation feels familiar and good, and what has happened feels right and effortless and just flows along. Figure 3.4 illustrates this statement.

Kahneman refers to his colleague Gilbert, who claims that the understanding of a statement begins with an attempt to believe the statement. – This also means that an overabundance of rationality hinders openness. Openness is one of the prerequisites for establishing resonant communication or creating novelty. Non-belief is the domain of rationality i.e. System 2 [8]:

The moral is significant: when System 2 is otherwise engaged, we will believe almost anything. System 1 is gullible and biased to believe, System 2 is in charge of doubting and unbelieving, but System 2 is sometimes busy, and often lazy.

From this point of view, good mood as one of the proponents of cognitive ease, can cause damage in certain situations.

According to Kahneman, but modified somewhat, we use the following definition of intuition: A situation, sensory perception, or a feeling gives us a clue. This clue gives us access to a variety of information stored in our memory, and this information is subsequently integrated into an intuitive answer. Intuition consists of recognizing and integrating the Recognized.

By unconsciously picking up specific clues from among all possible stimuli, and then using this clue to access specific mental information, a tendency (bias) or distortion can very easily push us in a certain direction. The type of information available depends on the cognitive ease with which this information is accessible. So, it is not the "right" information that comes into consciousness, but information that is most easily accessible. They will not necessarily coincide.

Figure 3.5 illustrates the relationship between System 1 and System 2 and shows their most important characteristics:

The central message of Kahneman and Tversky's work is that both System 1 (intuition) and System 2 (rationality) are needed in order to act under uncertainty. Kahneman and Tversky provide scientific insights as to how intuition and rationality work and give recommendations on how to combine both to achieve stable and "good" choices. This was presented by Kahneman in the very accessible book "Thinking, Fast and Slow" [8].

Figure 3.6 shows the central aspects of the "world of distortions". In the center of the figure is the node "Bias (distortion)" with its essential triggers: priming, anchoring, cognitive ease and WYSIATI (What you see is all there is). The biases lead to different manifestations of errors and illusions, as well as of heuristics.

In the following, we will explain the mechanisms behind priming, anchoring and WYSIATI (the importance of cognitive ease has already been explained above) and elaborate on their importance for project work.

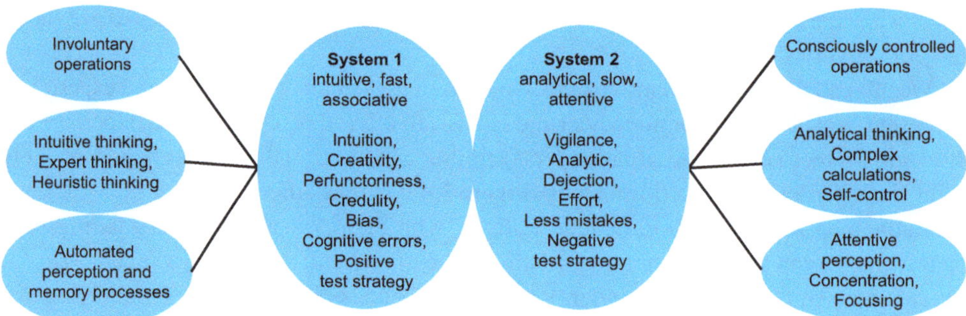

Fig. 3.5 System 1 and System 2

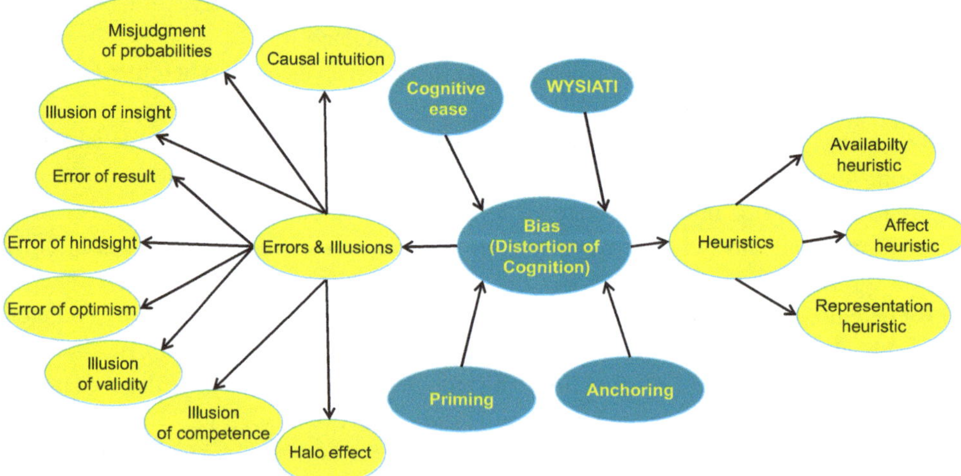

Fig. 3.6 The world of distortions

We will describe the heuristics through project work examples and limit our analysis of the resulting errors and illusions to the two cases of misjudgment of probabilities and the Halo effect.

Priming Priming is the canalization of thoughts in a particular direction. This canalization or steering of our thoughts can have "negative" as well as "positive" effects.

In the example of the "flying bank notes on the screen saver" the effect is probably regarded as negative, because the subjects became more selfish. In a project team meeting, agendas, project plans, central statements or images evoke, to different degrees, a strong priming effect. It is therefore important that project managers in particular are aware of the continuous presence of priming, how it works and how it can be used or influenced. If, for example, a new product, new process or new IT system is to be created in a workshop, the first person to present their ideas creates a priming effect. For this reason, it may be useful for all participants to put their ideas and/or images on paper independently of each other. Kahneman calls this countermeasure "decorrelating". The team members' mental models therefore, do not experience common canalization initially. At the beginning, especially in the context of an innovative project, it is always necessary to "decorrelate", as the possible solution space should be fully explored without early canalization. It is only when a possible solution emerges from the deliberate evolution of ideas that it is useful to establish this solution as an objective. Following this, the priming effect can be very beneficial: an image that embodies the solution, paves the way for the mind and secures orientation. Correct project naming is therefore of similar importance. Even the cognitive ease with which a name is perceived by stakeholders has an influence on the acceptance of the project, and this should not be underestimated. The name should be chosen in such a way that it opens up associations linked to the purpose of the project. The associations then pave the way again for further information on the project.

The creativity technique named after Walt Disney [19] uses priming in two ways: Each person in a creativity team adopts different roles (dreamers, realists, critics) while simultaneously changing rooms. Thus the dreamer role (here predominantly System 1) only takes place when a person is in the dream room, which is creatively designed to embody the role. Role and room have a priming effect on team members. The change from one room to the next both negates and enforces new priming. As we know, Walt Disney was very successful with priming strategy.

For example, a company emphasizes its value of "open communication" through spatial structure and setting up large offices. With this spatial structure "open communication" is "primed". Very often, however, at the same time, what is overlooked, is that this form of priming can counteract the value propositions of individual employees and is therefore counterproductive.[8]

Anchoring Anchoring has a very close affinity with the priming effect. An anchor represents a reference point to which we mentally "relate". Kahneman and Tversky identified a form of anchoring for both System 1 and 2.

[8] Priming thus acts as an order parameter and can counteract the values of employees whose mutual appreciation acts as a control parameter.

An example of an anchor in System 2: If for an uncertain quantity, e.g. the amount of work necessary for a task in hand, a value is estimated, this value then represents an anchor for System 2 (the rationality). Even subsequent adjustments to this value are no longer unrestricted, since you are no longer moving too far away from the anchor. Many project leaders are aware of this effect, and the more tired System 2 is, the more pronounced it is.

An example of an anchor in System 1: The project manager would like to hear from one of the team members how much effort a task requires. The team member hesitates before answering. The project leader becomes impatient and asks whether the upcoming task is easier or more difficult than the previous one. Due to how the question is now phrased, the question on the current effort is now linked associatively to the efforts of a previous task. There is canalization in the direction of the previous task. Associations (events, settings, emotions, etc.) associated with the previous task are activated in System 1 and are able to act as an anchor. The stronger the emotions involved with the previous task, the stronger the anchoring will be, in either a positive or negative way. In the case of positive experiences, effort is probably underestimated (the estimate is optimistic), with negative experiences, it will be overestimated (the estimate is rather pessimistic).

WYSIATI The WYSIATI (What you see is all there is) principle and its consequences occur in all kinds of biases. System 1 tends to use only information which is easily accessible by the mind. Because of this, the conclusions drawn often fall victim to considerable bias. Here are some examples:

The team member who is primarily familiar with their own work and thus overestimates it both in terms of quantity and quality.

The project manager, who has finished their last project with flying colors, and is selected by the company management for the next project, although they do not have the necessary expertise (see also Halo effect).

Technology that is regarded as good because it is constantly presented in public or celebrities represent it, even though it carries considerable risks or evokes consequential damage.

A project leader who thinks their project is complex, because they have members in the project team with whom they do not have a good rapport.

Heuristics Heuristic, or a rule of thumb, is a simple procedure that helps us to find adequate, though often imperfect, answers to difficult questions. A difficult question is usually unconsciously replaced by a simpler question, which is answered in turn. We often believe that we have answered a difficult question. In fact, we have merely answered a simple question, which can be arbitrarily distinct from the actual question. According to the mechanism by which the actual question is replaced by a simpler question, we distinguish between different types of heuristics: affect heuristics, availability heuristics, or representation heuristics. Since heuristics plays an important role in problem solving in projects, we have summarized some examples from project work in Table 3.1:

Table 3.1 Heuristics

Target question	Heuristic question	Type	Peril	Countermeasure
Is the partner/colleague the right choice for the task?	Do I like the partner/colleague?	Affect	The most likeable, rather than the best candidate is chosen	Establishment of a criteria catalog, use of a personality model
What are the risks you can see of using this technology?	Do I feel this technology is valuable and good?	Affect	Technology, which is considered valuable and well-known on the basis of existing information (WYSIATI) is used.	Use questionnaire, ante mortem method: "Imagine, we are a year in the future. We have implemented the plan in its current version. The result was a disaster. Please take 5 min to write a short story on this catastrophe." [8].
Have I done everything to make the project a success?	How much extra time and effort have I already put into the project?	Availability	Own past effort is over-weighted and energy is lost.	Visualization of results so far and associated expenses, as well as the necessary results and related expenses.
How likely is it that the project will be successfully completed within the appointed time frame?	Have I successfully completed all previous projects?	Availability	The risks in the project are not seen, as the project is associated with previous, successfully completed projects.	Based on external perspectives, request a project reference type for the project.
Is the designated project manager the right choice?	Do they make a good impression with their appearance? (e.g. the candidate has a distinctive face and/or is an extrovert, which is associated with assertiveness)	Representation	A project leader is appointed to the project who does not fit the requirements, resulting in a high level of unnecessary risks being taken.	Establishment of a criteria catalog, use of a personality model.
How complex is the project?	How much effort have I estimated is required for the project?	Representation	The risks inherent in the project are not transparent and there is no learning effect when dealing with complexity.	Use of project type models and project complexity models.

Heuristics can certainly provide a valuable contribution to decision-making when there is uncertainty. It is important to become aware of the potential dangers that can result from heuristic replacement mechanisms in a project.

We propose the following strategies:

- Use of simple criteria for answering the actual target questions. This prevents heuristic questions from creeping in unnoticed. The use of simple personality models will lead to the activation of System 2, the same applies to the use of cultural models in the choice of partner organization or the use of models for project typification for the assessment of project characteristics and associated risks.
- Conscious recruiting of team members with a pronounced System 2 and appreciation of the contributions of these team members. Since these team members are rather introverted in our experience, this requires corresponding awareness on behalf of the project leader.
- Conscious use of project outsiders' view: This could mean that information about individuals or similar projects is collected or that the project will be subjected to a "review" by an independent body.

These strategies counteract the WYSIATI principle, postulated by Kahneman (What you see is all there is – principle). System 1 tends to almost completely neglect the quality and quantity of information and simply takes what is easily available mentally. The above strategies support System 2, so System 1 cannot present what is mentally easily accessible too quickly.

Halo Effect This bias means that characteristics are attributed to an individual or an object, which they do not have. For example, an expert, who has been successful in other projects, is used in a project, although they may lack the technical background. The Halo effect means that you do not notice this, or only notice much later once details emerge that make the lack of technical background apparent. But, by then it may be too late for a cost-effective correction of a project. The Halo effect results from an overvaluation of effects associated with an individual: The sportsman or woman breaking records, the entrepreneur who leads a company to prestige, and the project manager who successfully completes a difficult project. In all these cases, so-called regression to the mean is forgotten. Our intuition underestimates the "power of the big numbers": when a project leader carries out many projects, there is a high level of probability that they will be mediocre on average. It is a fact that our intuition almost always underestimates the importance of statistics and gives priority to specifics.

Misjudgment of Events with Low and High Probability A central result of Kahneman and Tversky's research is the so-called new decision theory. It states, among other things, that when making decisions, people do not behave rational, as economic standard theory assumes. The following statements apply to negative events (reverse statements apply to positive events): Negative events with low probability are more highly appreciated in their (perceived) probability of occurrence (risk aversion), and negative events with high probability have a lower level of appreciation in their (perceived) probability of occurrence (risk appeal). Kahneman speaks of the possibility effect and the certainty effect.

Risk seeking (possibility effect): Let us assume there is a very high probability (95%) of not completing a project within time frame, budget and quality. In this case, a significantly lower decision weight (for example, 75%) is assumed for the failure of the project. Emotionally, this means that there is the hope of avoiding losses and successfully completing the project. Decision-makers behave much more adventurously than they usually would and "burn" money by not stopping the project in time.

Risk aversion (certainty effect): Here on the contrary, we assume that there is a low probability (5%) of unsuccessfully completing the project. The decision weights are higher, 15%. Decision-makers are (unconsciously) afraid of unsuccessfully completing the project and behave in a more risk-averse manner than the actual probability figure suggests they should.

In this way, errors can not only relate to the entire project, but to all the individual aspects of a project.

Rare or even insignificant events can lead, for example, to the following behavior:

- Low probabilities for risks, associated for example with tasks, lead to risk aversion; the tasks are postponed or circumvented
- Low probabilities of possibilities associated with tasks lead to risk-taking; the task is gladly carried out, expenses explode and the tasks turn out to be pointless.

Events that are almost certain, lead to the following behavior, for example:

- High probabilities of risks, associated with tasks, lead to risk-taking; the tasks are carried out with higher expenses and relatively poor quality.
- High probabilities of chances, associated with tasks, lead to disproportionate risk aversion; the execution of tasks are postponed or circumvented.

The significance of these misjudgments for project work is only recognizable when we realize that the assessment of risk and chance, as well as low and high probabilities, is in the eye of the beholder: if there are different views in the stakeholder circle that are not transparent, this again increases the overall risk within the project.

3.3 Regulation of Complexity Through Targeted Interconnectedness and Self-Organization

…Léone had been hit because of the butterfly in Brazil whose wing she had seen move. …
Finding a butterfly's wing in a settlement of two thousand inhabitants was trickier than the proverbial needle in the haystack. Something which had never seemed insurmountable to Adamsberg: all you had to do was burn down the haystack and you'd find the needle.

Fred Vargas, The Ghost Riders of Ordebec,
English by Siân Reynolds

We have seen above that key to regulation is using the appropriate variety of the model that regulates a complex system.

In principle, teams have the possibility of developing more variety than individuals would. So if the team's variety can be increased with a specific goal in mind, then a team is a (smart) regulator for complexity: A prerequisite for this is, however, that the team's variety develops in line with the tasks, i.e. value-creating complexity is created within the team: In other words, the whole is more than the sum of its parts.

In this section, we will follow our discussion of self-organization in [20].

Self-organization is nature's means of creating, through networking, systems with high levels of variety. Self-organization is interesting, because it leads to emergent macro-states in complex systems, which are characterized by a few order parameters. With technology, laser light is an example of an emergent macro-state. In nature, our mind, which is formed from neurons, is an example of the formation of an enormous number of macro-states (see the section "Philosophy of Complexity" and Edelman and Tononi's approach, which is mentioned there, describes consciousness through networking and integration of information). In a team, a Collective Mind is an emergent macro-state. In human society, culture is viewed as a emergent macro-state. Emergent macro-states go beyond the properties of the elements of a system. They belong to the core of complexity and constitute the properties of the system as a whole.

If the emergent macro-state Collective Mind is present, we are talking about a high performance team. The total performance is then greater than the numerical sum of all individual performances of the team members. Figure 3.7 shows team performance, according to Tuckman [21, 22], once for a "normal team" and once for a high performance team with a Collective Mind.

In the "Forming" phase, the team gets to know each other: If the team members do not know each other, or are not yet well acquainted with one another, they will be cautious in

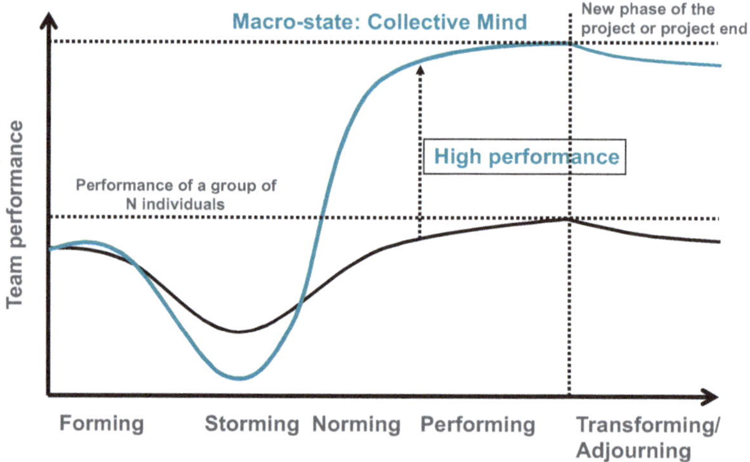

Fig. 3.7 Collective Mind and high performance

their remarks. The task and its solution are probably unknown to all of them. A "normal" team and a high performance team differ only insignificantly in this phase of team performance. Team performance level is usually below the performance level of the sum of the individuals: In a team consisting of 5 people, we speak of team performance of 5, or better 500%, if each team member exhibits 1 or 100% of their performance. In the above figure, this performance level has not been achieved and is clearly below 5. Team members would be more powerful if they were to work alone instead of in the team. Different personalities, different interests and different understandings of the task and of the conceived solutions, result in frictional losses.

It is in the "Storming" phase that these frictional losses become apparent. There are misunderstandings of technical aspects, different interests become obvious, behaviors are rejected and in a non-appreciative manner, "bad" personality traits are assigned to one other. Team performance continues to fall. In a team, which really ought to show high performance later on, "conflict" is deliberately brought about. This leading into "conflict" requires an experienced guide, because the "conflict" must not go so far that irreversible personal distress occurs. The "conflict" creates stress, which everyone should perceive as controllable. As we shall see in the section "Transformation Management", the course of team performance according to Tuckman, corresponds to the system of transformation work by Virginia Satir. For Satir, the storming phase is one of mental chaos: interventions lead to irritations that start new neural interconnectedness in the brain and thus learning begins. In a team process, team members mutually evoke these irritations.

In order to find a way out of the "Storming" phase and to enter the "Norming" phase, an experienced guide is necessary. Through targeted interventions, the integration of the mental models of all team members is induced. It is necessary for each team member to integrate their ideas with the "new whole". Self-referentiality plays an important role here, at the individual level (micro-tier) and also at the system level (macro-tier). We will see below how self-referentiality is supported: the core element is transparency, that is, the visualization of ideas and basic assumptions, and the joint creation of an integrated "Gestalt" in the form of the solution. From here onwards, the "normal" team and the high-performance team begin to differ substantially.

In a high performance team, the "Performing" phase exhibits a level of team performance, which may exceed the sum of individual performances. In Haken and Schiepek's book of self-organization in psychology [23], team performance is measured by quantifying the proportion of mirroring in communication. Mirroring means that facts are repeated in a communication with communication partners using the same words, or that facial expressions and body postures are aligned while communicating. NLP refers to such communication as resonant communication, with calibration. Calibration is the conscious perception of the communication partner with all five senses and associating perceived behavior to perceived emotional patterns. For example, voice pitch and facial expressions of a communication partner in the team are associated with excitement or anger. After calibrating, trustworthy communication based on empathy takes place, which is called "rapport". During rapport, among other things, mirroring communication elements are used.

This "measurement" of the quality of communication and associated team performance can be carried out by the team members themselves and made visible to all.

The visualization of the assessment of team performance by all team members in or after a workshop is a very good way of encouraging self-referentiality. By encouraging self-referentiality by this or similar means (this can also be carried out by using tools such as Lessons Learned, Retrospective or Review) team performance is sustainably increased.

Figure 3.8 shows two examples of team performance according to team member perception. Team performance perceived per team member is recorded over the course of the workshop: Once for a team of five team members and once for a second team of six team members. As can be seen from these figures, the observed courses are very different, at times contradictory. Experience has shown that perceived courses in team performance correlate with the learning of a greater level of attentiveness and mindfulness, and correspond better to external observation. At the same time, with the matching of the curve shapes, there is an associated increase in team performance.

It was also observed that team members with similar personality profiles tended to have similar perceptions, and that different personality profiles showed very different patterns in perception. For example, personalities with pronounced goal orientation (MBTI preference "judging") tend to distort the results (see above): This is reflected in the fact that at the end of the workshop, a high level of team performance is indicated – even if this was not the case, provided the results of the workshop are in accordance with the goal of this individual. In Fig. 3.8, this pattern can be seen in the left-hand section: At the end of the workshop, one person appreciates the team performance significantly more than other team members. On the right-hand side, there is another pattern: One team member believes that team performance has continuously increased and that team performance was even higher than the performance level of the sum of the individuals shortly after the start of the workshop. The majority of the remaining team members also

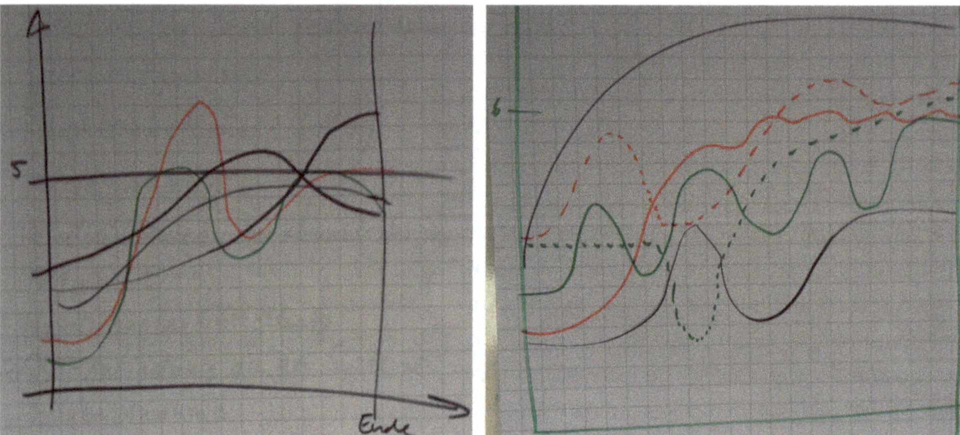

Fig. 3.8 Team performance, assessed by team members

experienced a positive trend during the workshop, but with obvious fluctuations. Team members with MBTI personality preference N (MBTI preference "Intuition") tended to overlook details and color the entire course in line with their preferences. In the present case, this led to bias with a positive attunement: Fluctuations in the course were not perceived as long as a large trend (the "big picture") was perceived moving upwards (see appendix "Fundamentals MBTI").

Let us turn to the question of how we can intentionally initiate social self-organization in a team so that a high performance team can develop.

Social self-organization is not understood here as a model of normative self-determination or as the right to "organize ourselves", "give ourselves our own rules" or "have no leadership". Social self-organization may include these, however, this is only an essential aspect if, along with these aspects emergent macro-states are generated.

The concept of self-organization and its associated emergence is discussed from various angles in scientific literature [24]. The only operationalized theory, known to us, that has been successfully applied to natural, psychological, and sociological emergent macro-states is Synergetics [23]. Along with this, we use a theory that is operationalizable and can be falsified in practice.

On the basis of eight natural-scientific principles, we will show how self-organization can emerge and how goal-oriented variety of a team can be shaped by means of the Collective Mind. These are very basic and directly applicable principles, which are core element of Synergetics. With these tools in mind, the prerequisites and emergence of self-organization can be understood and applied. It is not our intention to apply the (mathematical) formalism of Synergetics to project management. Instead, we show how the eight principles of self-organization determined by Haken and Schiepek can be applied to a team to create a Collective Mind.

Dealing with Complexity: Few Principles Are Sufficient

Priesberg and Ehrlich meet again in the elevator. Priesberg is still getting his teeth into complexity. "So far you have been able to convey to me one aspect or other. But, if I ask you now, how to regulate complexity and describe it in a nutshell, then you have to give in and tell me, right?" Ehrlich grins, "Now I have you where I want you: There are eight principles that regulate complexity, and I will explain them to you by the time this slow elevator has reached the top."

We speak of a self-organized team, if the following eight principles are applied and the team is developing an emergent macro-state, a Collective Mind, which is in line with the project goal. All principles should be applied equally; there is no explicitly specified order. The order of the principles listed below suggests, at most, a sequence for the application. During a team-building process, the principles should be applied several times in order to achieve a stable emergent macro-state. The application of the principles means the deliberate adjustment of the described setting- and control parameters. The formation of corresponding order parameters and emergent macro-states is a consequence of this.

Principle 1: Creation of Stable Setting Conditions Stable setting conditions (setting parameters) are necessary so that a team can develop attentiveness and mindfulness, as well as concentration and focus. Stable setting conditions include, among other things, regular meetings, introduction and adherence to "rituals", a stable team composition, and the explicit shielding of complexity through shielding in space and in time. Shielding in space and/or in time can take place by having all team members seated in a room where outside disturbances are prevented, or having a team meet up away from day-to-day operations. As we have already mentioned above, mental self-organization in the team arises through the integration of team members' mental models into a Collective Mind that is related to the task. Disturbances, especially in the building-phase of the Collective Mind, prevent mental networking. This is most easily illustrated using the example of a new team member. A new team member brings their own vision, identity, beliefs and values, abilities and behaviors, to the team, in short their personality. If the Collective Mind of the original team is not yet stable, it can be considerably irritated, in the worst case scenario, even destroyed. In the language of Luhmann, we could say that self-referential team structures cannot be formed [25]. Therefore, in the Collective Mind Method [26] we also state reference values for the new inclusion of team members, and also for the appropriate mix of personalities in the team. It is self-evident here that these stable setting conditions, which favor self-referentiality, do not set themselves up. It is the task of the team or project leader to ensure this necessary stability. The agile framework Scrum has introduced an explicit role for this, the Scrum Master, and in addition, introduced the product owner as a "gate keeper" to the complexity of the customer world (see also the chapter "Consequences for Management Systems").

Principle 2: Identification of Mental Structures In nature, this principle is represented by the fact that not every natural system exhibits self-organization. The ability of a system to create emergent macro-states is essentially determined by its individual elements, its agents. The agents must have the ability to interact with other agents in a specific way. In social systems, these agents are people, who adjust adaptively according to their environment. It is of central importance to know the individual personalities in the team sufficiently well, to be able to adapt to them and stay in line with the overall goal. This statement applies to all team members, but in particular, to the team or project leader. Our experience has shown that there are people who have the ability to intuitively appreciate the individual personalities of team members and communicate with them accordingly and mindfully. However, as a guideline for all team members, we recommend the use of the Dilts Pyramid in combination with models such as MBTI for temperament, or the Reiss model for motives and the conscious and structured treatment of personal values and beliefs (see the respective appendices "Fundamentals MBTI" and "Fundamentals Reiss Motive Profile"). Therefore, we do not recommend the application of agile values from the range of agile books to Scrum or Kanban, but instead we consider that appreciation of individual temperaments and the individual values of each team member is indispensable. It is only in this way that self-organization can develop. This is supported by modern research results in neuropsychiatry [27] and is connected to the topic of Neuroleadership [28] (see the corresponding appendix "Fundamentals Consistency Theory" and the section "Neuroleadership"). The task of a team leader is therefore to enable the team members to get to know each other's personalities. In principle 4 "Activation of emotional energy" we return to this again. The aim of the principle "identification of mental structures" is to create prerequisites so that the team members can enter in resonant communication. This means that the communication amplifies itself positively with regard to the task at hand. There should be appreciation for each other, openness to different ideas and suggestions, in short, one's own interests should be seen and weighted in the context of the "big picture". If mental blockages occur, the team will have appropriate tools (i.e. Dilts Pyramid, MBTI, Reiss Motive Profile, value and belief analysis) to make these blockages transparent to all and dissolve them.

Self-organization therefore needs leadership. This often appears to be a late insight, but is self-evident from the understanding of self-organization described here.

Principle 3: Creation of Meaning and Purpose In the natural examples of self-organization, this principle is directly connected with the fact that natural systems, through self-referentiality, form an emergent macro-state. Transferred to social systems this means that people perceive a "higher" meaning and purpose in the project tasks or in the organization's purpose. It is only in such cases that individuals seem willing to subordinate their current interests to this purpose or to adopt this purpose as part of their own professional activities. Creating meaning also implies, for every individual, bringing the current situation together with current, past and future mental representations. A person who is unable to reconcile the project goal with their current and future career plans will not support an emergent macro-state in the sense of the big picture. Based on this knowledge, in the Collective Mind Method [26] we have derived "design specifications" for the design of the project goal: The project goal is defined by a target hierarchy in such a way that the different temperaments of all team members are taken into account. This actively supports mental networking and the focusing on a common goal. It is of central importance that this target hierarchy is not defined by the project manager or by the product owner, but developed within the team itself by self-organization. In order to proactively support this mental self-organization, namely the creation of a target hierarchy consisting of a big picture, goal clustering and specific objectives, pictorial visualization of ideas is crucial. This visualization is a technical method for initiating the team's self-referentiality. Furthermore, other methods that support the five senses can also be used. In the language of the Dilts Pyramid, the top levels (vision, mission, belongingness and identity) are created for the temporary organization, called "project team". In the language of Synergetics and complexity theory, a top hierarchy of order parameters is formed, so that, individual behavior in particular, along with its manifold possibilities is aligned with the big picture and therefore subsequently limited.

Principle 4: Activation of Emotional Energies Without intrinsic motivation of the team members, the formation of the team's emergent macro-state is doomed. Everyone must be able to contribute with their needs and abilities. Needs, skills and challenges should ideally be balanced. If this is the case, Mihály Csíkszentmihályi says that an individual is in a flow state [29]. It is no easy matter to achieve this flow state at all times simultaneously, for all team members, but it is the specific task of the leader to work towards this ideal. This is a challenge every day, even for an experienced leader. This applies in particular to team formation at the beginning, with the aim of achieving an emergent macro-state. In the last two decades, neuroscience has gained considerable insights, and these findings have led to a new view of leadership, Neuroleadership (see also the corresponding following chapters, as well as the appendices "Fundamentals Consistency Theory" and "Fundamentals Reiss Motive Profile"). The findings in neuroscience suggest that human beings have four basic needs: the need for affection, the need for pleasure and avoidance of pain, the need for orientation and control, and the need for self-appreciation and self-protection. These basic needs vary from person to person. Everyone is constantly trying to balance their individual manifestations of their basic needs, while at the same time being active, i.e. interacting with their environment.

Let us take an example: A team member's basic need for pleasure and avoidance of pain is particularly well satisfied through the "curiosity" motive. It is very important for them to bring a new idea to the team or project. The leader is either unaware of or ignores this motive and assigns the team member to a project that could be seen as a routine project. The team member is left unchallenged and begins to get bored. A flow state fails to ensue. The team member is dissatisfied, and in the worst case internal dismissal occurs. Leadership could have easily prevented this if they had become aware of the situation and the employee had been deployed to another project, which emphasizes novelty. However, this is highly unlikely to occur if the company does not have a pioneering or entrepreneurial spirit, but instead attaches importance to safety and order. In this case, the employee in our example would have no choice other than to leave the company.

It is hypothesized that motives have a genetic origin or are at least formed in very early childhood. Values are driving forces, more likely to be cultivated as a result of later experience. Thus, a team member appreciates the value "health" significantly more, if in the course of their life, illness has raised the value "health" in the hierarchy of values. Both values and motives take over the function of order parameters in our lives, they "absorb" complexity. A team member with the "health" value may no longer be willing to extend their working time beyond 8 h, but might prefer a sporting activity. Another team member, who places importance on the "transparency" value, will be willing to make their work results available to other team members. In this way, a value for each respective person "absorbs" complexity, because the value restricts behavior possibilities. However, motives and values can also increase complexity, namely, when the motives and values of an individual are in conflict with the motives and values of another team member. For instance, one team member is driven by the need for orientation and control, whilst another team member loves their "autonomy". In this case, two different behaviors are likely to occur which exhibit occurrences that are detrimental to the team or the project goal. There will be similar consequences when the motives and values of an employee are contrary to the values of the organization. Motives and values influence our entire behavior and our overall communication. Attentive and mindful adjustment to the motives and values of the other acts as a control parameter.[9] It is therefore necessary for team members, but above all for respective leadership, to recognize the values behind behavior and verbal communication. The basis for mental connectedness, as mentioned above, is resonant communication that recognizes and resonantly amplifies the motives and values of each team member in daily communication. This is much more than merely agreeing a range of values as rules, for the team. Agreements, and especially rules based on values cannot replace individual, mutual appreciation of team members.

Principle 5: Admit Fluctuations To find new ideas, it is necessary to open up to creativity. This can also be positively influenced by appropriate selection of team members with suitable personalities. In the section on intuition, we have already shown how the admission of new ideas (especially at the beginning of a project) is necessary in order to both create something new and initiate mental networking, and in addition, to find and illuminate risks. It often happens that a project leader interrupts a discussion for their own agenda. In hindsight, it might turn out that a topic, which would have been of enormous importance for the project, was either abandoned or was not covered. The style of leadership must therefore, not only adapt to the phase in the project, but also to the maturity level of the Collective Mind. Team members who repeatedly describe similar situations, are an indication of a not yet existing, or at least unstable Collective Mind. It is in the first third of a project, in particular, where the idea of the goal and the way to it are concretized, and where it is necessary to allow the opportunity for open discussion. At a suitable point, the discussion must be intercepted and reviewed in terms of quality and outcomes. For those people who have a pronounced need for orientation and control, this allowance of fluctuation can actually result in painful reactions: "Where is our agenda?… we are losing time… why is it taking so long ….". Getting through this tension can be even more challenging if the organization places a high level of importance on the values of order and safety. It is then even more important that the related communication patterns are addressed transparently. Additionally, it is necessary, at the beginning of a project or team formation, to communicate the basic principles of resonant communication.

Principle 6: Resonant Interventions As we have described in the previous principles, resonant communication is one of the keys to the formation of the Collective Mind as an emergent macro-state. Consequently, there are specific requirements for leadership interventions, since such interventions are always accompanied by the label of power. This results in either a desired or undesired reinforcement of the intervention. The Collective Mind as an emergent macro-state is a collective, mental entity with a high level of

[9] Parameters such as values can occur as control parameters or as order parameters. For an individual, motives, values and basic assumptions operate as order parameters, as well as values and basic assumptions, which an organization decides on, and acts on as order parameters. In a team or an organization, mutual appreciation of individual motives, values and basic assumptions by team members, however, operate as control parameters.

complexity. However, this also means that the smallest intervention can lead to enormous effects. The intervention must therefore fit the cognitive-emotional state of the team. Interventions in a complex system always mean that the effect is unpredictable. As shown in Dörners's book on thinking in complex situations [4], the effects of interventions often only occur after a delay, so it cannot always be verified as to whether the intervention was a step into the right direction or not. An example may illustrate this: A team has a new idea, and would like to check customer needs by means of a prototype. The executive sees this as a waste of time and makes his feelings known. At first, nothing much happens; team members only display anger initially. However, team members slowly begin to self-regulate their drive and similar experiences with management are pigeonholed as the same management pattern. A new rule of thumb is formed "proactive action with new ideas is undesirable". A "negative" order parameter has been created, which blocks the formation of a Collective Mind.

Principle 7: Enable Deliberate Symmetry Breaking As part of the solution, several new ideas are presented, they are checked and either accepted or rejected. After a specific length of time, indicated by the occurrence of a sufficiently stable Collective Mind, it becomes necessary, for the sake of further work that the search for new is no longer attached any importance: All alternative ideas and solution concepts are deliberately pushed back, and only those preferred by the team are highlighted as solutions and selected for further elaboration. Not all solutions are of equal importance anymore, but by team selection, a specific solution is chosen as the best one: The symmetry in the space of the initially equivalent ideas has been broken and only one solution from among many possibilities is pursued.

In order to create new structures, it is not only variety that is necessary, but also the compulsion of the system to select certain states, so that one or several states are dominant. Niklas Luhmann expressed this idea with one of his central statements on the system theory of social systems: "complexity is called … compulsion of selection"[10] [25]. High Variety ("freedom by control parameter") in conjunction with compulsion of selection ("focus by order parameter") is a central characteristic of self-organized systems.

Principle 8: Induce Stabilization If a solution has been chosen by means of selection compulsion from a range of possibilities and is found to be viable, this also involves a radical change in leadership style: Openness to innovation, the fluctuation phase, is now followed by a phase of stability. The spectrum of values has changed. It is one of the central tasks of leadership to deliberately shape this transition and make it transparent to the team. The process of "admitting fluctuation and subsequent stabilization" can certainly occur

[10] Translation by the authors.

more than once during a project. In the Agile Framework Scrum, the sprint has the function of carrying out this transition in a controlled manner. In sprint planning, customer requirements are (newly) recorded, and from the beginning of the sprint onwards, selected stories are processed. External complexity is blocked and the need for order and control is supported within the team. In the Collective Mind Method [26], we have pointed out that it would be useful to support this phase in the project by slowly introducing personalities into the team that will contribute to stability.

The above form of self-organization is just one of several possibilities: On the one hand, we have used the theory of Synergetics as a basis, and on the other hand, we have interpreted the eight principles of Synergetics using the Collective Mind. Based on our experience, we assume that this described version of social self-organization in teams is an appropriate representation.

The formation of macro-states by means of self-organization is a universal phenomenon, which occurs in various forms. In the section "Philosophy of Complexity", we have already mentioned the importance of this universality. Here, we will discuss the various aspects of the system elements expressed in the principles, their interaction and related setting, control and order parameters. In the section "En Route to Complexity", we gave examples of natural, technical, socio-technical and social systems with self-organization. Let us take a closer look at the LASER example: The system elements of the LASER system are atoms or molecules that are excited by irradiated energy (control parameters) to form pairs of particles. A special mirror arrangement (setting parameter) ensures self-referencing, "enslaving" all other excitation and only a specific excitation is amplified – the symmetry is broken – and thus established as an order parameter.

In the example of the team, the system elements are the team members. The setting parameters are ensured by the above-described setting conditions, as well as by means of mirroring by visualization with the Collective Mind scheme. The set-up of the control parameter is achieved by mutual appreciation via resonant communication. The formation of the order parameter "Collective Mind" is initiated by resonant communication based on the Collective Mind scheme and leads to a common perspective on the project goal, which is supported by everyone.

In the section "En Route to Complexity", as an additional socio-technical system we have referred to a company's multi-project management design, based on the "Theory of Constraints" and the Critical Chain Approach by Goldratt [30]. Goldratt's basic ideas are certainly systemic, but it is little known that Goldratt, as a physicist, drew a connection to self-organization. We have seen above that in systems where too many connections exist, they become inflexible. When a system becomes inflexible, it is no longer able to cultivate self-organization. This is even more so if the determinant system element, the bottleneck, is affected. The number of concurrently running projects using the same resources act as control parameters. In the example of team self-organization, too many tasks handled by one person at the same time (Agile Management speaks of work-in-progress) also act as control parameters that completely "drown" the attentiveness and mindfulness control parameters. If the team members "juggle too many balls at once", whether by being involved in several projects or by working on several tasks simultaneously in a project, any cultivation of attentiveness and mindfulness is doomed to fail, since it cannot even occur.

A multi-project management organization must also be set up first using the control parameter "Work-in-Progress" so that an order parameter can be built up and create self-organization for the projects' systems. In the Critical Chain Approach, this is represented by the fever curve of the multi-project management system. The fever curve is a 2-dimensional figure of buffer usage and prognosticated project progress. The buffer, and therefore buffer usage, is determined on the basis of the belief that "estimates are 50% overestimated". The exact size of the buffer is not important in our view. What is important is that an explicit risk indicator is deployed in addition to project progress, and that the buffer counteracts by generating additional pressure. It is also important to ensure the generation of self-referentiality by means of uniform visualized information, i.e. a "big picture" of the multi-project management system. The system elements of this system are the projects, even if control takes place via project teams. The whole system is controlled by a few rules that determine which project to prioritize throughout the system according to the fever curve. This form of self-organization only works if the individual projects themselves do not exhibit any limiting factors. If a project team does not perform well, the fever curve makes it visible, but team performance problems cannot be solved on this basis. For this, it is necessary to enter the individual project. If the Work-in-Progress problems of the entire system are resolved, problems that go beyond "simple" resource issues have to be solved at the individual project level. This is where self-organization of the team becomes crucial as described above. The performance of the overall system is therefore also determined by the performance of individual projects. The performance of individual projects is a setting parameter for the performance of the projects' overall system. This is why, the project's Critical Chain Approach focuses on the introduction of projects: projects which are not sufficiently well-prepared are not started. The highest level of system performance is subsequently only attained when multi-project management and individual project management are guided by means of self-organization.

Genesis Workshop

The Genesis Workshop is a team game format to provide, within a short time, an introduction to self-organization. As a prerequisite, the fundamentals of the Collective Mind and the model of self-organization need to be known. Ideally, the workshop room should contain creative game material.

The team should consist of at least 4 team members, and a maximum of 10.

Each team member takes one of the following four roles. These roles follow the creativity technique created by Walt Disney [19] (Table 3.2):

Table 3.2 The Genesis roles

Dreamer, Visionary	The dreamer is very visionary and creative; they surprise the team with many new and creative ideas
Realist, Manager	The realist checks new ideas for feasibility and plans their implementation
Critic, Reviewer	The critic looks for errors and potential for improvement in the ideas, plans and their implementation
Observer, Consultant	The observer observes the team and advises the team

The task of the team is to find a creative solution for the following task within a 1.5-h workshop:

As a team in TecFiction company, develop a machine or a device based on high-tech components (nano, biotec, light guides, …). For potential customers, the machine or the device should be highly sought after. Any kind or form of machine or device is allowed.

Rules of the game and instructions:

- Each role should be filled by at least one person from the team. Use your MBTI skills to team up. Visualize your team with the MBTI house (see Fig. 7.5). (Depending on the learning goal, it may be useful to specify that the dreamer/visionary role is filled by the person most suitable for this purpose, and so on.)
- Work with the Collective Mind scheme to identify and present the goal.
- Please note that the workshop style/leadership style should change during the course of your workshop.
- After an hour, the team will present its results:
  - The presentation should take a maximum of 5 min.
  - The presentation should contain statements on the task and the solution (target hierarchy).
  - The presentation should contain the key information of solution and implementation (basic principles, basic elements of the solution and the creation process).
  - Present (at least) one key piece of information in more detail.
  - Observers should report on three key observations.
  - The team is completely free to design its presentation; any available object in the room can be used.

3.4 Regulation of Complexity Through Organizational Setting, Control and Order Parameters

Be not afraid; the isle is full of noises,
Sounds and sweet airs, that give delight, and hurt not.
Sometimes a thousand twangling instruments
Will hum about mine ears; …

William Shakespeare, The Tempest

Sometimes Even One's Own Temperament Has Undesirable Effects

Priesberg is looking for Ehrlich in his office. "I sort of like my new boss, because he does not go into all the details with me. I even have more freedom than before." Ehrlich interrupts him: "Are his values and principles now in line with those of your colleagues and your environment?" "He's getting there slowly...," answers Priesberg with a little hesitation, "but he has a good relationship with our customers, he has a communicative personality. Although this probably undermines the role of the employees, who are supposed to be the ones who talk with the customers, right?"

Ehrlich slaps the palm of his hand against his forehead: "Well, then your boss should take a refresher course in self-organization, because he is blocking his employees by his tactical behavior."

Priesberg interrupts: "I think I know what you are trying to say: Only when the values of both the environment and the boss are in harmony, can a stable order develop – and the boss should also stick to them."

Ehrlich is quite taken aback: "It looks like our conversations have proven themselves effective and you are ready to continue on the path to complexity. I have heard from a reliable source that the new research manager, Herbert Freschi, is about to approach you concerning a very delicate matter, but I will not reveal any more just now." Ehrlich bites his lips, as he realizes he has unintentionally confused Priesberg and perhaps unintentionally set up a blocking control parameter for him. He is no longer able to keep his curiosity under control: His true personality has caught him out again. Ehrlich disappears quickly, leaving the baffled Priesberg behind, mouth open, reminiscent of Edvard Munch's Scream…

After reviewing the principles of complexity regulation for individuals and teams in the previous sections, in this section we turn to larger organizations, such as departments, companies, and organizational networks.

Each organization embodies organized complexity. Each organization also contains different elements (people) that are aggregated into new elements (teams, departments, business units). All levels of aggregation, both stable and temporary, can be modeled in their organized complexity using the Dilts Pyramid (see appendix "Fundamentals Dilts Pyramid"): Organizations also act within a context and show behavior within this context, which is more or less appropriate to the context. The behavior is described in the organization through processes in which individuals are involved as agents. The organization needs specific skills for this purpose. These skills express themselves either through physical products or specific types of service. Individuals and groups bring their values into the organization, and in the course of organizational life, these are interwoven with experiences to become either successes or failures. These experiences are then the basis for the formation of basic assumptions and beliefs. If the behavior of the founder of an organization is characterized by flexibility, and this flexibility is rewarded by success, this behavior is transformed into the values of the organization. At the same time, this is also linked to the formation of beliefs that fit these values and the situation in which the success or failure occurred. For example, ad hoc measures may be positively reflected in sales and profits, which results in the basic assumption or rule of thumb that "flexible behavior means that every customer call is answered immediately". In addition, of course, these values and basic assumptions are shaped by a comprehension of identity, belonging, vision and mission. If the organization understands itself as a service provider that meets the requirements of the customer, or as a mechanical engineer with a solid, self-engineered product, then this will probably lead to different expressions of the understanding of flexibility: In the case of the identity concept "service provider", the previously discussed basic assumption is likely to be complemented as follows: "Flexible behavior means that every customer call is accepted immediately and customer requirements are met carefully". In the case of the identity concept "mechanical engineer", the extension of the basic assumption might be as follows: "Flexible behavior means that every customer call is accepted immediately and the customer is convinced by the special characteristics of our own product".

Since teams, departments and companies all have different backgrounds of experience and have also been influenced by different personalities, the structure of the Dilts Pyramids is usually very heterogeneous in slightly larger organizations. This generally leads to each team, each department each business unit, and of course the company itself, being described by another Dilts Pyramid.

Let's take an example with the help of Fig. 3.9: Let us assume that a department consists of three teams. The department is described by an embedding Dilts Pyramid. Each of the teams within the department is described by another team-specific form of the Dilts Pyramid and, of course, each team member is modeled by their individual manifestation of the Dilts Pyramid. In our model, we describe each team member in terms of their private contribution, team contribution, and departmental contribution with respect to the logical levels of the pyramid. This diversity is expressed in the different roles everyone takes depending on the context: In the team, individual-specific manifestations of the Dilts Pyramid "Team" are expressed, and private and department-specific parts are pushed into the background. This also shows the enormous significance of shielding in space and time, as outlined above. If shielding fails, no team-specific Dilts Pyramid can develop and the team will not exist as a team. This model also makes it very clear that the logical levels of the various contexts of individual, team and department must match each other. If, for example, the private value proposition, the value proposition in the team, and the value proposition of the department are too far apart, internal and external conflicts are preprogrammed.

To further illustrate our example, let us assume that the three teams belong to an IT department. Each team works very successfully on a specific IT solution segment. One team carries out their projects through Scrum, another team prefers classical project management, and a third team swears by Critical Chain Project Management. The department has won a new customer who has purchased an IT solution package to be composed of the individual solutions of these three teams. Everyone is thrilled at the new order, but already in the first team meeting, communication difficulties have arisen. The Scrum team is reluctant to set a defined time and cost frame and would like to have the requirements in the form of user stories. The team working with classical project management, swears by the

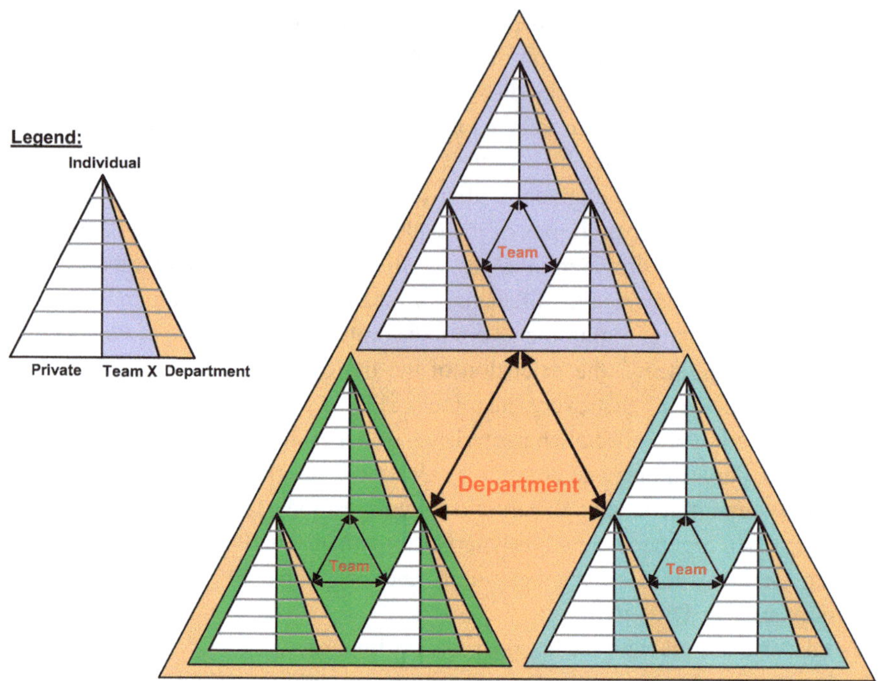

Fig. 3.9 An organization (department) with three teams, displayed via the Dilts Pyramid

creation of a requirement specification and clean planning. The team that wants to use Critical Chain Project Management, disapproves of all of this and swears by agile steering using project progress and buffer size. In joint project meetings, the customer also notes different approaches and inconsistencies in the behavior of the team as a whole. After a while, sloppiness is visible in the envisaged solution. Subsequently, the customer decides to cancel the project.

What happened? The department was unable to adapt to the new situation caused by the new constellation demanded by the project. The past successes of the three teams led to them believe that they are right and to the development of values and beliefs that fit their respective success. Unconsciously, rules of thumb developed as belief systems and evoked corresponding mental distortions, for example. "Scrum is good", "Without a plan it's just chaos" or "Critical Chain always helps". These distortions are usually visible when the context in which they were created no longer exists.

This is where organizational transformation work takes place: It begins with the recognition of values and beliefs and the context in which they fit. For each of the teams it is necessary to determine the exact context in which their behavior led to success. Subsequently the corresponding values and beliefs are worked out and reference to the context established. Thus, the values and beliefs of the three teams become visible as "as well as" order parameters. It is necessary to create new common values and beliefs for the three teams. Through this process of transformation, the three teams are freed in their actions, and the organization increases its ability to adapt and is able to regulate complexity. In the section "Transformation Management" we will return to the topic of transformation work.

Based on this example, we note that the mindset, the set of all manifestations an individual or organization has on all logical levels, exhibits different role characteristics: In our example, we have those of private individuals, team members, and members of the department. Behaviors and skills will vary according to these different roles and their related contexts. This is normally not a major problem. If the values and basic assumptions show a great level of differences in the different roles, dysfunctional effects are to be expected in the medium term. An employee may be more adventurous in their private life, yet place an (adapted) emphasis on safety within the team. If much larger differences occur, detrimental effects may occur in the medium term. If a team or a department attaches great importance to safety, such an employee will be put off by the organization. They will not be able to reach their maximum performance and the organization will judge their as unsuitable for the organization. If this dysfunctional trend continues, in the upper levels of the Dilts Pyramid, serious consequences, including health consequences, are to be expected.

This dysfunctionality described for individuals, caused by different mindsets, can also exist in teams, between teams or in other organizational units. Thus, as described above, the mindset of one team may not match the mindset of another team completely or in part. Regulation of complexity in team composition therefore, requires disclosure of the mindset, a corresponding dispute in the team and the alignment of mindsets on the upper levels.

On this basis, how can regulations of complexity in an organizational context be established? In order to establish regulations of complexity in an organization successfully, attention needs to be paid to the neurological levels of an organizational mindset, to ascertain on the one hand that they are aligned perfectly, and on the other hand that they have kept their adaptability with respect to the organizational environment.

Reorientation of a company through the "Vision" and "Mission" levels is a frequently practiced method used by managing directors and board members, which only really succeeds in a few cases. "Vision" and "Mission" are targeted here as hierarchy of order parameters. The problem here is that order parameters should come from the system itself, rather than be defined by a few people. A further requirement is that the manifestations of all logical level should match. If, therefore, the levels "Identity" and "Values and basic assumptions", which have potentially developed over many years of experience in a company, do not fit with "Vision" and "Mission", friction arises. The result is that complexity is not regulated, but increased.

The levels "Identity" and "Values and basic assumptions" are used either consciously or unconsciously in addition to "Vision" and "Mission" as additional order parameters. Since individual motives, as well as the values and basic assumptions of the organizational employees directly influence their abilities and their behavior, the use of the "Identity" and "Values and basic assumptions" levels as organizational order parameters can have an immediate dysfunctional effect. In the section "Regulation of Complexity through Targeted Interconnectedness and Self-Organization", we therefore placed emphasis on the statement that organizational values should either not be used, or should be used with great care, as order parameters. If organizational values are "merely" used as control parameters, this also means that individual appreciation of employees' motives and values is in the foreground, rather than organizational "order". This is because values that are used as order parameters can have an "enslaving" effect. The smaller the organization, the more immediate the effect. Therefore, care should be taken that values as order parameter do not conflict with the control parameter function. As a consequence, this means that all leadership parameters – setting, control and order parameters – need to be tuned: As we have already seen in the previous sections, the wrong choice of order parameters can end up making any form of complexity regulation impossible.

With respect to the environment of a project, let us turn to the adaptability of neurological levels: For management of individuals or organizations, knowledge of respective mindsets is of enormous importance for stakeholder management. Only through an understanding of the interaction of all mindsets, is it possible to identify friction and blockades and to intervene specifically in a stakeholder group.

We distinguish between two different cases where interventions are applied:

- Case 1: The complexity of the environment can be influenced, because the environment consists of a network of organizations, which are at least partly in the influential sphere of an (temporary) organization: For example, a project involving different areas of a company as stakeholders. Another example is a department that interacts with other departments.
- Case 2: The complexity of the environment can be very affected very little or not at all. An example of this is the market or society, whose complexity cannot be regulated by the organization.

Since it is not always a good idea to regulate complexity by shielding in space and time, as this would mean excluding the environment in extreme cases, it is necessary in both of the above cases to continually adapt to the environment. At the same time, this adaptation must not jeopardize or destroy the organization, whether it be the company, company department or project. In recent literature, this requirement has been represented by two image concepts: An organization should be an anti-fragile system [31] or a fluid organization [32]. At this point, we will not delve further into the term "Fluid Organization", since in our chapter "Consequences for Management Systems", we have included a separate section named "Fluid Organization".

The fragility of a system is based largely on non-linear effects. Let us look at an example: A new IT system was introduced very rigidly by means of a project. In order to achieve stability, the fluctuations of the social system (objections, questions, and requests to change anything) are not allowed. After a certain period of time, the first flock of rejections emerges, and a little later, a flurry of unpleasantness spreads among users. Through the networking of user mindsets, a so-called self-organized criticality [33] emerges, which has ultimately erupted from the smallest trigger into massive rejection. From this example, a measure for the absorption of complexity can be deduced: Fluctuations are the expression of different mindsets. As we have seen in the section on self-organization, stability cannot be acquired without fluctuations. It is therefore necessary to allow fluctuations, accompanied with targeted interventions in the system. These interventions can only be professional if the organizational mindset of the project and the mindset of the environment are known and their interactions are heeded. In this example, anti-fragility of the "project" system means to have the necessary flexibility to allow fluctuations.

Let us look at another example. Fragility is generated when the system does not allow options: In a company, new project management (PM) processes are introduced. There is a belief that adhering to the processes and their associated roles and artifacts is essential for good project management. Shortly after the PM processes have been introduced, the context of the company changes. Implementing projects according to PM processes brings the company in conflict with its customers; and customer orientation dwindles. From this example, we can deduce that it is necessary to allow several options on the process level, i.e. to allow several methods of process, and only preset cornerstones for the execution of the process. In this example, anti-fragility of the "PM processes" system means having the necessary flexibility to maintain the "customer orientation" value and not fix specific behavior.

As we can see from these two examples, suppression of fluctuations and options in behavior initially blocks complexity by narrowing behavior. However, regulations are even more devastating in the long run. Whereas allowing fluctuations and options acts as a control parameter that gives stakeholders room for the necessary mental adaptation work.

In order to allow fluctuations and options in an organization, while at the same time limiting diversity, to avoid uncontrolled growth, behavior is limited and stabilized by more abstract guidelines (vision, mission, values, beliefs). The consequences of complexity are regulated by a hierarchy of well-tuned order parameters according to the Dilts Pyramid model.

This insight can be applied to the two cases outlined above: "Case 1: Environment can be influenced to a certain extent" and "Case 2: Environment cannot be influenced". If the behavior of the organization and its members are not regulated by means of behavioral rules, but instead by higher order parameters such as vision, mission, values and beliefs, the result will be a much broader range of behavior. This makes an organization (also an individual) anti-fragile: The organization recognizes order parameters in its environment and can check and adjust its own order parameters to influence its behavior. Its behavioral repertoire, becomes substantially larger on the one hand, and on the other, much more adaptive. This is the basic idea of organizational transformation work in Business NLP [5] and how we treat it in the section "Transformation Management". We would also like to mention at this point that we regard agility (Agile Organizations) and fluidity (Fluid Organizations) as a means to preserve the anti-fragility of organizations. We will return to this issue in the chapter "Consequences for Management Systems".

Literature

1. Ashby WR (1957) An introduction to cybernetics. Chapman & Hall, London
2. Conant RC, Ashby WR (1970) Every good regulator, must be a model of that system. Int J Syst Sci 1(2):89–97, London
3. Scholten DL (2010) The Good-Regulator Project. Every good key must be model of the lock it opens. https://en.wikipedia.org/wiki/Good_regulator. Accessed 14 Feb 2016
4. Dörner D (2011) Die Logik des Misslingens: Strategisches Denken in komplexen Situationen. rororo Verlag, Kindle Version
5. Brinkmann M (ed) Besser mit Business NLP. DVNLP e.V., Berlin
6. Mohl A (2010) Der große Zauberlehrling. Das NLP-Arbeitsbuch für Lernende und Anwender. Teil 1 und Teil 2. Junfermann Verlag
7. Oswald A, Müller W (2015) Trainingsunterlagen: Agiles Projekt Management 4.0. GPM Deutsche Gesellschaft für Projektmanagement e.V., Nuremberg, unveröffentlicht
8. Kahneman D (2010) Schnelles Denken, Langsames Denken. Siedler Verlag, Munich
9. Gigerenzer G (2014) Risiko: Wie man die richtigen Entscheidungen trifft. C. Bertelsmann Verlag, Kindle Version
10. Gigerenzer G (2008) Bauchentscheidungen: Die Intelligenz des Unbewussten und die Macht der Intuition. Wilhelm Goldmann Verlag, Munich
11. Edelman GM, Tononi G (2013) Consciousness: how matter becomes imagination. Penguin Press Science, Kindle Version
12. Oswald A, Mourgue d'Algue H, Merk G (2014) Fundamental requirements for a framework of adapted theories for project work. In: Rietiker S, Wagner R (eds) Theory meets practice in projects. GPM Deutsche Gesellschaft für Projektmanagement e.V., Nuremberg, pp 52–65
13. Böhle F, Kuhlmey A, Müller F, Siermann P (2014) Uncertainty in projects – new demands and approaches. In: Rietiker S, Wagner R (eds) Theory meets practice in projects. GPM Deutsche Gesellschaft für Projektmanagement e.V., Nuremberg, pp 163–176
14. Wikipedia (2014) Viable system model von Stafford Beer. http://de.wikipedia.org/wiki/Viable_System_Model. Accessed 3 Dec 2014
15. Dilts RB (1992) Einstein: Geniale Denkstrukturen und Neurolinguistisches Programmieren. Junfermann Verlag, Paderborn
16. Beck H (2013) Biologie des Geistesblitzes: speed up your mind! Springer Spektrum, Berlin/Heidelberg. Kindle Version
17. Damasio AR (2001) Ich fühle also bin ich. List Verlag, Munich
18. Oswald A, Köhler J (2013) Schnelles und langsames Denken in Projekten: Zur Beherrschung von Unsicherheit in Projekten, Teil 1. projektManagement aktuell 5:30–36
19. Wikipedia (2015) Walt-Disney-Methode. https://de.wikipedia.org/wiki/Walt-Disney-Methode. Accessed 15 Jan 2016
20. Oswald A, Köhler J, Schmitt R (2015) Wertschaffende Komplexität und Selbstorganisation. projektManagement aktuell 4:45–51
21. Wikipedia (2015) Phasenmodell nach Tuckman. https://de.wikipedia.org/wiki/Teambildung#Phasenmodell_nach_Tuckman. Accessed 3 June 2015
22. Prudix D, Goerner M (2009) Teamarbeit in Kompetenzbasiertes Projektmanagement (PM 3). GPM Deutsche Gesellschaft für Projektmanagement e.V., Nuremberg, pp 351–430
23. Haken H, Schiepek G (2010) Synergetik in der Psychologie: Selbstorganisation verstehen und gestalten. Hogrefe-Verlag, Göttingen
24. Greve J, Schnabel A (eds) (2011) Emergenz: Zur Analyse und Erklärung komplexer Strukturen. suhrkamp taschenbuch wissenschaft, Berlin
25. Luhmann N (1987) Soziale Systeme, 1st edn. Suhrkamp Verlag, Frankfurt am Main

26. Köhler J, Oswald A (2009) Die Collective Mind Methode. Springer, Heidelberg
27. Grawe K (2004) Neuropsychotherapie. Hogrefe, Göttingen
28. Peters T, Ghadiri A (2013) Neuroleadership – Grundlagen, Konzepte, Beispiele: Erkenntnisse der Neurowissenschaften für die Mitarbeiterführung. Springer Gabler, Wiesbaden
29. Wikipedia (2014) Flow (Psychologie). http://de.wikipedia.org/wiki/Flow_%28Psychologie%29. Accessed 5 Dec 2014
30. Techt U (2015) Projects that Flow: Projekte in kürzerer Zeit. ibidem, Kindle Version
31. Taleb NN (2012) Antifragile. Things that gain from disorder. Random House, New York. Kindle Version
32. Weßels D (2014) Zukunft der Wissens- und Projektarbeit: Neue Organisationsformen in vernetzten Welten. Symposion Verlag, Dusseldorf
33. Bak P (2014) How nature works: the science of self-organized criticality. Copernicus, Kindle Version

4 Leadership in Complex Social Systems

'Léo goes on about the butterfly's wing', Èmeri continued. 'She says the important thing is to spot the moment it moves. ...'

...

Léone had been hit because of the butterfly in Brazil whose wing she had seen move.

Fred Vargas, The Ghost Riders of Ordebec, English by Siân Reynolds

In this chapter, we will elaborate on the basic principles of complex systems, which were developed in the previous chapters, with models for social techniques.

In the section "Fundamentals of Leadership", we begin by describing leadership in complex social systems by using the Dilts Pyramid and we concretize this by using an example of value-oriented leadership.

Complex systems demand continuous learning and the development of meta-competency, to perceive systems, i.e., from an outside perspective using a meta-position. Leadership means, to a great extent, being able to adopt this meta-position. This aspect of leadership will be discussed in the section "Learning and Meta-Competency".

In the following two sections we apply value-oriented leadership and demonstrate how resonant communication can be derived from this. We demonstrate the enormous importance of self-reflectivity and the importance of knowledge of one's own personality traits for the elaboration of leadership.

In the section "Neuroleadership" we introduce consequences, derived from neurosciences, for human-oriented leadership. We demonstrate the connection to the model of consciousness and "Spiral Dynamics" culture. With the integration of these different models, an integral view of leadership will become visible.

A. Oswald et al., *Project Management at the Edge of Chaos*,
https://doi.org/10.1007/978-3-662-48261-2_4

In the section "Team Leadership, Intuition and Biases", we demonstrate how a leader can utilize known mechanisms from the section "Interplay of Intuition and Rationality" to develop innovative solutions without mental biases in a team.

All this results into two further sections; both are devoted to stakeholder management. In the first of these two sections, also named "Stakeholder Management", we show how the tools introduced so far are used in implementing effective stakeholder management through social networks and personality trait conceptualizations. In the section "Interaction Between Micro- and Macro-tier", we extend this perspective by incorporating organizations with their cultures and the interaction of individuals and culture. Thereby, leadership and stakeholder management are combined in a new and practical manner.

This chapter concludes with the section "Transformation Management", in which we raise the question as to which types of organizational change work and which types of interventions can be derived from them in order to stimulate such change work. The section concludes with four intervention examples on transformation management.

4.1 Fundamentals of Leadership

... in one voyage
Did Glaribel her husband find at Tunis,
And Ferdinand, her brother, found a wife
Where he himself was lost, Prospero his dukedom
In a poor isle, and all of us ourselves
When no man was his own.

William Shakespeare, The Tempest

In the Collective Mind Method [1] we elaborated Malik's management model as a leadership model and explained the principles, tasks and tools of leadership. In the language of the Dilts Pyramid, we have described the values and beliefs (principles), capabilities (expressed in terms of the ability to use tools) and behavior (tasks).

Here, we go noticeably beyond this approach and understand leadership as a meta-competency that is able to regulate and organize complexity.

En Route to an Agile Organization, Part 1

Heiner Priesberg ends the call and turns to a colleague at the next table: "Well that was a strange conversation with Herbert Freschi, head of research. I am supposed to think about creating a directive for a specific measurement procedure. It relates to a multi-step process to ensure the quality of active ingredients produced in the research department for test purposes."

"I'm sure he will tell you more about it tomorrow," his colleague assures him. However, Priesberg has a bad feeling.

The next morning, Priesberg enters Herbert Freschi's office, who gets straight to the point: "So, I've heard you are an excellent biotechnologist and that you are also on good terms with the project enthusiast Ehrlich. Therefore, I believe you are ideal for the following task: In our three test departments, active ingredients from the research department are tested. Testing has obtained completely different results for the same substances, dependent on the test department. Of course, this will not do. It is therefore necessary to establish a directive which ensures that measurement procedures reach the same conclusions, regardless of the department in which they were carried out. Provided, of course, that the same test substances are used. The directive should be designed to guide employees into taking the right steps. Do you see where I'm coming from? I am pretty sure that all three departments have developed different value judgements, which we will now have to balance." Freschi looks at Priesberg briefly, then looks at the clock, "You know that this is a complex problem right? Unfortunately, I have to go to my next appointment, but I'm sure you'll do just fine."

Detail-obsessed Priesberg remained upset for a while after the conversation with Mr. Freschi and questioned: "Why does Freschi not explain in detail how I should create and implement the directive?" He stares arrogantly into space: "A directive is not complex, but complicated." He continues to draw himself deeper into the issue: "It is a great example for complicatedness: It contains comprehensible instructions and is detailed. Complex would be if it contained many interactive dependencies, but how can that be so with text? I don't think Freschi quite understood the matter. Typical management: They just bandy terms around. Today the 'in' word is 'complex', tomorrow it will be something else." Priesberg is still thinking. Suddenly he smiles to himself: "Actually, in itself, it is quite simple: as a biotechnologist, and from a professional point of view, I probably know best what needs to be ensured with the directive. I'll simply write it down and the directive is ready. Everyone can then read and understand it themselves." Priesberg returns his thoughts to the here and now, and affirms: "What a simple job to create a *complicated* directive!"

This definition of leadership contains three central aspects: Meta-competency, regulation of complexity and organization of complexity.

We speak of meta-competency, because it relates to the competence to perceive sovereignly the manifestations of one's own personality and that of communication partners, in as many facets as possible, i.e. "walking mentally up and down the Dilts Pyramid". This also includes recognizing manifestations of an organization's Dilts Pyramid – for instance recognizing a temporary organization's mindset, a project's mindset, and its interaction with the mindset of the members of an organization and the interaction of the organization's mindsets. For this purpose, it is not only necessary to be able to "walk mentally up and down the Dilts Pyramid", but it is also necessary to step out of the system of which one is part of and look at this system from the outside, from a meta-position.

Based on the previous chapter, leadership is largely recognizing value-destroying and value-creating complexity and regulating it (passively) or organizing it (actively). As a means to this end, in the previous chapter we discussed shielding in space and time, the use of interconnected theories and models and the use of related intuition, the elaboration of organizational setting, control and order parameters, and the organization of complexity by self-organization.

We will now go on to explain this distinctly extended understanding of leadership. We begin in Fig. 4.1 considering a system of two communicating individuals and model them using individual Dilts Pyramids, i.e. by their individual mindsets.

The individuals communicating could be a team leader and a team member, or an employee speaking to their boss, or two colleagues looking for a solution to a problem. In all these cases, the respective individuals are embedded in a systemic context. Even if both individuals are currently in the same situation, their whole context is different, because they are acting out different roles, have a different personal background, and may be located in different departments of the company. In a specific situation, in their specific personal context, these two individuals will demonstrate behavior, which is determined by their own specific capabilities. Capabilities include outward visible abilities of linguistic communication and body language, the ability to perceive emotions in oneself and others, and the ability to solve problems or professional expertise. Emotional abilities include the recognition of one's own feelings, accepting these as emotions and either transporting them to the outside or keeping them to oneself.

As experience shows, our behavior is determined by our capabilities and is also regulated by our values and basic assumptions. We deliberately use the verb "regulate" to express the fact that a few values and basic assumptions regulate the range of our behavior by means of our skills. The value "trust" serves as an example.

If an individual has confidence in themselves, i.e. self-confidence, this individual is able to adapt their own behavior without a comprehensive review of a new situation. As we all know, stage fright is an expression of situation-related uncertainty, and gives rise to different behaviors such as "repeatedly playing out the situation" or "obtaining acknowledgment from others". In short, lack of self-confidence generates different behaviors, thus evoking complexity.

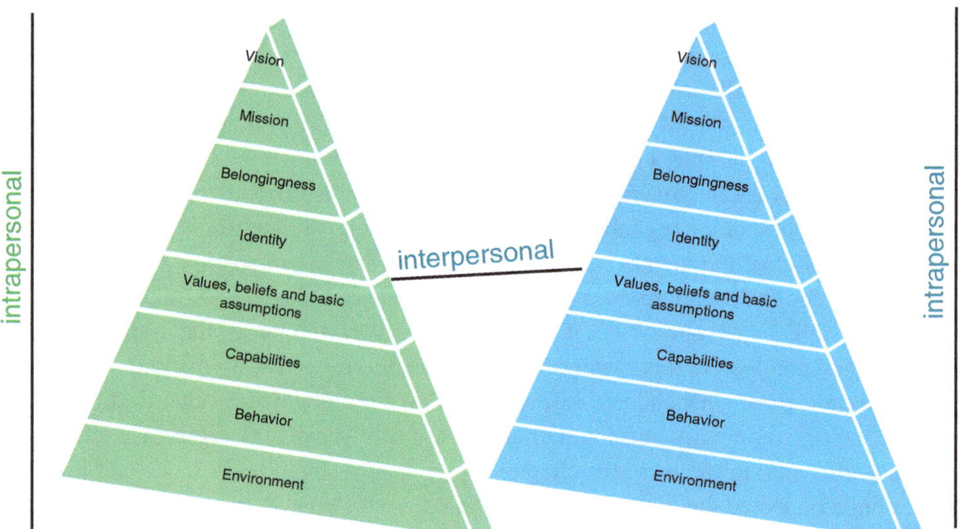

Fig. 4.1 Self-leadership and leadership

It is only when self-confidence returns through an experienced situation (for example, the presentation is accepted by the audience) that uncertainty is reduced and behavior stabilizes.

When communicating with another person, lack of mutual can lead to a number of correlated behavior patterns.[1] In projects this can very quickly lead to over-control with excess suspicion and/or an attempt to somehow regulate the related complexity by the use of extensive contract documents. As we know, this will rarely succeed, as contracts can not replace mutual trust and furthermore, can induce even more complex behavior in all parties involved (refer to the example in this section, "En Route to an Agile Organization").

The higher we move toward the top of the Dilts Pyramid, fewer and fewer order parameters regulate our behavior. With the level of "identity" our personal identity is expressed. When, on the values and basic assumptions level we ask: "What is important to me?" and "What do I believe?", we are asking on the identity level: "Who am I?", "Which temperament do I have?", "What drives me?", "Which needs do I have?", "Which motives do I have?" Temperament can be expressed using the MBTI model and needs by Grawe's model of basic needs and associated motives according to Reiss (also refer to the appendices for these models).

The knowledge of these models facilitates cognitive, and later emotional access to our own identity. Based on these models, we are able to detect and structure patterns of our own behavior; they help to attain consciousness about ourselves. Moreover, it is from this, that the term meta-competency is derived. It is an ability to look at oneself, to look from the outside at one's own personality, which is modeled by the entire personality-specific manifestation of the Dilts Pyramid.

If I am unaware of my abilities, or do not know what values and basic assumptions have shaped my experience, or I am not familiar with which temperament glasses I am looking at the world with, or which motives drive me, I am an "outstanding" source of value-destroying complexity. Because my behavior is less intelligent in a given situation (the context), less adaptive and therefore "non-compatible" behavior is generated. This in turn, also leads to a smaller possibility for adaptation in the communication partner. Value-destroying complexity increases.

Sometimes the level of belongingness is neglected in representations of the Dilts Pyramid. However, we recommend maintaining this level, as identity manifests through belongingness.

When I feel part of the project management community, it gives a specific framework to my personality. I identify with the understanding of professionalism, which is inherent to (some) project managers. There is a similar effect when I say "I am a veterinarian" or "I am a member of the local football club". In addition to self-understanding, belongingness also expresses organizational affiliation.

[1] "Correlated behavior patterns" are behaviors that arise from interactions in which no linear cause-effect relationships occur.

In the case of an organization, one of the ways this is expressed is by affiliation to the branch of industry: "The company belongs to automotive suppliers". This reflects a certain similarity in corporate culture, processes, structures and strategies, as well as in the employees who work for these companies (we speaks of a certain background or "barnyard").

Now let us visit the top levels "Mission" and "Vision". They are connected to the questions: "Why are we here?" and "Where do we want to go?" or "How could we contribute to a greater whole?" Two of the authors of this book feel that they belong to the community of physicists. Another of the authors perceives himself as belonging to the community of engineers. Both communities map the perceived reality into models. Our common mission therefore is through models, the social techniques described here, to better understand and describe complex social systems in order to deliberately regulate complexity and use it to generate value.

In the section above "Regulation of Complexity through Targeted Interconnectedness and Self-Organization", these levels are strongly linked to the term "Sense". Sense exists when we are able to bring our current roles into balance with our past and future. In addition, the top logical levels of the Dilts Pyramid play a central role here. Leadership first off means self-leadership, i.e. to become aware of what personal preferences on the various levels of the Dilts Pyramid are connected to one's own behavior. If this awareness exists, then the higher levels can be applied for targeted influencing of our own behavior. Self-leadership therefore is intrapersonal connectedness of the neurological levels of our own personality. The goal of self-leadership should be to perceive one's own personality through meta-positions, have as few blind spots as possible and regulate it accordingly.

In a further step, leadership means enabling other individuals who are being guided, to perceive their personality at all levels of the Dilts Pyramid, i.e. to effect an adaptive interconnectedness of the personalities in a team or an organization, as well as to guide this interconnectedness towards a common goal. We have seen above that this exactly what happens within the setting of self-organization.

To understand this better, we recommend carrying out a self-assessment of your own temperament according to the MBTI and assessing your values and motives. Regarding the MBTI, please refer to the Collective Mind Method [1] book and Keirsey's questionnaire [2], which is available on the Internet.

To facilitate the identification of values, in Fig. 4.2 we have compiled typical values and motives

You will find your personal values and motives by asking "What is important to me?", "How do I spend my time", "What do I spend my money on?" Please try to identify your most important values and motives and rank them according to relevance. When you have done this, please create a "Personality-flower" by adding your MBTI temperament as shown in Fig. 4.3.

Keeping your Personality-flower in mind, now try to ask yourself, what behaviors you exhibit according to the respective values and what behaviors you expect from others who

Fig. 4.2 Values and motives (blue) according to the Reiss Motive Profile

Love,	Adventure,	Trust,
Altruism,	Independence,	Family,
Attractiveness,	Fun,	Harmony,
Team-orientation,	Helpfulness,	Responsibility,
Health,	Growth,	Beauty,
Freedom,	Goal-Orientation,	Physical Activity,
Power,	Creativity,	Curiosity,
Passion,	Romance,	Development,
Fighting,	Obedience,	Order,
Vitality,	Integrity,	Saving,
Success,	Eating,	Ambition,
Acceptance,	Openess,	Status,
Respect,	Safety,	Energy,
Honor,	Helpfulness,	Flexibility,
Idealism,	Influence,	Tranquility,
Professionality,	Social contact,	Vengeance, ...

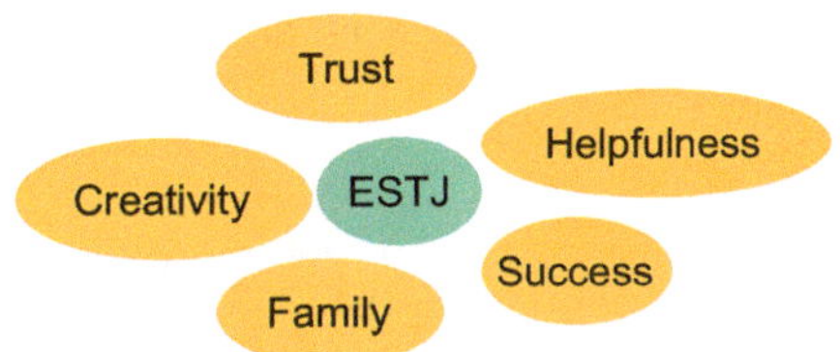

Fig. 4.3 Personality-flower

appreciate your motives/values and your temperament. Based on our example above, this could be the result (please see Table 4.1):

Several different behaviors may be exhibited or expected to belong to a value or motive. These may differ from one individual to the next. Two people with the same values or motives, might differ in behavior or may expect different behavior. In NLP we speak of evaluating behavior of so-called complex equivalences[2] that has not necessarily occurred, but is expected. "My boss does not agree to me driving home at 5 pm, so he does not respect me as a person." Or "the stakeholders constantly submit change requests, so they are trying to torpedo me". From these examples it can be seen that complex equivalences lead to more social complexity and that they also hide their origin due to their structure, i.e. it can be quite difficult to reveal them in communication, because only in rare cases are people are aware of the connections.

Before applying these insights to leadership, using the Dilts Pyramid, we look at what kind of leadership styles exist. Figure 4.4 illustrates, by using the Dilts Pyramid, the most important leadership styles found in literature.

[2] Complex equivalences are directly related to motivational schemes and are caused by incongruities: If the motivational scheme is not satisfied, it is expressed by the complex equivalence (you see the appendix "Fundamentals Consistency Theory").

Table 4.1 Values and behavior patterns

Value or motive	Shown behavior	Expected behavior
Family	I leave my office punctually at 5 pm so that I can spend time with my children.	My manager agrees that I can leave my office at 5 pm.
Success	My projects are always within budget.	I do not expect my Stakeholders to submit change requests.
Helpfulness	I am happy to help my colleagues.	I would hope that in difficult situations my colleagues would not stab me in the back.
Trust	I do not check the work results of my team members.	I do not expect my team members to scrutinize my project plans.
Creativity	I do not always want to work on the same type of project.	I would hope that the projects are not long-running and that my manager gives me at least two different projects per year.

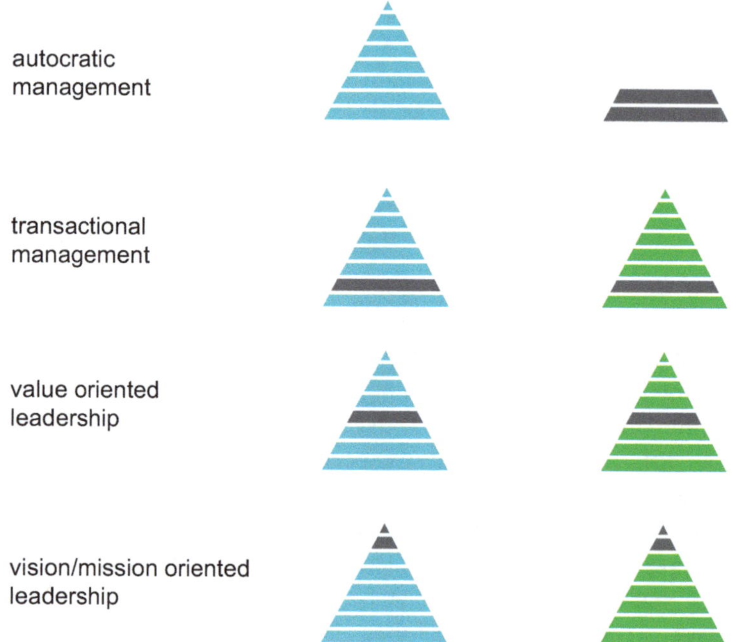

Fig. 4.4 Leadership styles

An autocratic leadership style is characterized by complete indifference of the leader (left in the picture) towards the individual/s they are leading. The leader is only interested that the person they are leading exhibits expected behavior in each situation. This means that with required behavior, the person being led is not existent, they are merely a resource whose behavior is utilized. For this reason, we have not included the upper right levels of the Dilts Pyramid.

Transactional leadership style focuses on behavior. The leader and those being led agree on specific behavior: Those being led are compensated (salary, bonus, etc.). So a transaction is agreed.

Value oriented leadership style focuses on the level of values. Here, however, caution is advised. What might happen is that the leader tries to "impose" their values or the values of his organization on the person they are leading. This is not our understanding. We have already pointed out that values and motives can function as either control or order parameters. If the leader appreciates the values and motives of those they are leading, these values take over the function of control parameters.

Value oriented management will sometimes be understood in terms of the values of an organization serving to enable leadership. In this case, values act as order parameters: The setting of values by management or by the culture of an organization is primarily for the values to bring about "order". This is certainly a legitimate interest for an organization, but can also be very subtle "enslavement", especially if the values and motives are represented as the values and motives of an organization, but in fact are only the values and motives of a minority (e.g. management, peer group).

Vision/mission oriented leadership style focuses on the two top levels of the Dilts Pyramid. It might be the most difficult leadership style, but can develop an incredible leverage. In many cases however, only minimal effect is shown from this leadership style, because it is necessary to involve the lower levels of the Dilts Pyramid when using the vision and mission as order parameter. A vision and mission, which do not address the values and basic assumptions of those involved, show only minimal effects. The same of course applies in an organization:

New visions and missions, which do not take into account existing values and basic assumptions of organizational culture, remain "stuck" and will not be implemented. For example, if an existing organization is to be transformed with a new vision and mission for a service provider, the current mindset has to be taken into account. If the organization has the mindset of a product manufacturer, it is unlikely that the values and basic assumptions (as part of the mindset) will match the values and basic assumptions of the service provider. Vision and mission fail miserably in the central part of the Dilts Pyramid of the organization, which is sometimes wrongly assigned to the middle management.

We will focus on the value oriented and vision/mission oriented leadership: Value oriented leadership is able to utilize values and basic assumptions in a twofold manner: As order parameter and as control parameter. Values act as order parameters, if values are intended to "enslave" behavior and to regulate complexity by these.

The mutual appreciation of team member values or appreciation by the team leader, however, acts as a control parameter. An atmosphere arises in which team members can develop sufficiently. In addition, conditions are created in which they can collectively develop order parameters. This could be common values or a common vision and mission (i.e. Collective Mind). Ideally, an order parameter hierarchy of vision and mission is formed, including common values and beliefs systems.

On this basis, we can extend our above understanding of leadership: In self-organized organizations, leadership means to ensure, using appropriate setting and control parameters, the forming of order parameters in line with the purpose of the organization.

4.2 Learning and Meta-Competency

There, sir, stop:
Let us not burthen our remembrance' with
A heaviness that's gone.

William Shakespeare, The Tempest

En Route to an Agile Organization, Part 2

Before he starts to write the directive, Priesberg asks Dorothy O'Brian, who is responsible for electronic archiving, to look for all the rules for carrying out measurement procedures. He exclaims enthusiastically: "I intend to create a super directive. Everything is precisely specified!"

After some thought Dorothy gives replies: "What will you do if the details of the measurement procedure change? Would you rework the policy? And what about training, or instructing those colleagues who will be affected? The details of the text are well and good, but those who will be affected should really be involved. Is this all not more likely to be complex?", she hesitates, then continues: "I have often heard you talk to Tobias Ehrlich on project management topics. I am aware that his mind ticks differently than most of us. Before you create such detailed rules, I would talk to him about it."

Priesberg replies in a slightly offended tone: "I have to be able to implement a project on my own without the help of third parties. And, furthermore, this directive is quite a complicated problem, yet complex is only a trend word that is currently on everyone's lips." Dorothy shrugs her shoulders and leaves Priesbergs' office, wondering at his lack of willingness to learn. Nevertheless, Priesberg reflects: "We recently discussed the Dilts Pyramid for dealing with social complexity. We most certainly have to draft the directive in such a way that the values and basic assumptions of all three departments are taken into account – but wait a second: The implementation relates to a scientific and technical procedure. Through the language of science, I can still communicate to everyone what we need to take into account with respect to personal values and basic assumptions. In science and technology, everything is strictly logical and unambiguous."

He begins to write and finish everything within one night. In the end, there are fifteen pages, regulating even the minutest detail. The next morning, he plans to discuss the guideline step by step with the three representatives, the key stakeholders from the department. Priesberg drives home contented, totally unaware of the storm that is brewing.

As we have already pointed out in the previous chapter, leadership is closely linked with the terms learning and competence. If an environment is characterized by lack of straightforwardness and unpredictability, it will be inevitable that new situations will have to be faced and adapted to.

Adjustment will take place by learning and by the ability to look at our own actions from different perspectives and perceptions positions. The ability to remove ourselves from a situation and to take a meta-cognition position, to pursue completely new methods of action from gained knowledge, is a meta-competency that we combine directly with leadership.

The systems theorist Gregory Bateson distinguishes four stages of learning ("learning stages") that Dilts, DeLoizier and Bacon Dilts assigned in [3] to the logical levels of the Dilts Pyramid (Fig. 4.5).

We have illustrated the stages of learning according to Bateson, at the hands of a fictitious project manager Paul Little who develops into an excellent project leader.

Learning Zero

In the "Learning zero" stage there is no learning, only a repertoire of behavior patterns that are always applied, no matter what the situation is.

Example: Project manager Paul Little repeatedly uses the same method to carry out a project.

Learning I

In stage I, the patterns of behavior will be adapted or supplemented to a limited extent.

Example: Project manager Paul Little has heard of agile methods and enriches his planning with these methods. He has not been made aware of the environment in which agile methods are useful and when they are not.

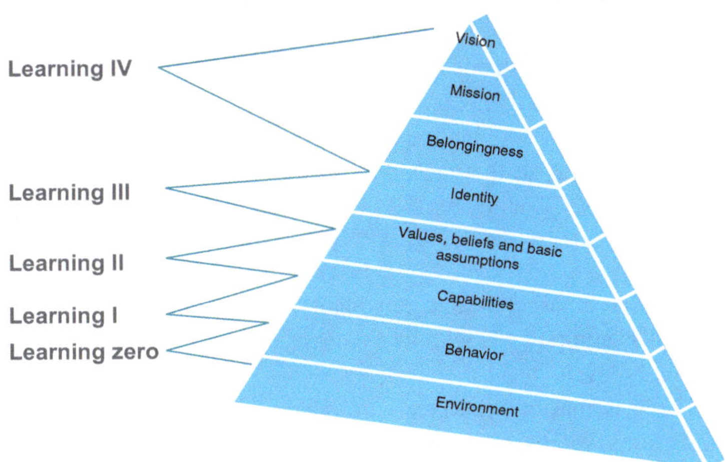

Fig. 4.5 Dilts Pyramid and the stages of learning

Learning II

In stage II, a weak meta-competency begins to emerge. The environment itself and the meaningful behavior for this environment, as well as relevant capabilities are recognized. Behavior patterns from other situations are also implemented, extending capabilities accordingly.

Example: Project manager Paul Little realizes that his planning method will bear better results in one environment, but will worse results in another. In future, Paul Little will first check the project environment and select his planning method accordingly.

Learning III

In stage III, meta-competency begins to emerge. Our personality is perceived in all its facets. We become aware of our temperament and our motives and the resulting behavior. Beliefs, basic assumptions and values are recognized and our own belief system and values can be profoundly changed and fully replaced, if necessary. It is then that we have gained the ability to recognize completely new patterns in other people and to both model and apply them.

Example: Project manager Paul recognizes the values and beliefs which have led him to select his project planning method. He checks to see which principles, in which project environment, are meaningful, and then selects his project methodology accordingly.

Learning IV

In stage IV, meta-competency has fully formed. We are able to "step out" of ourselves as a system, but are also able to "step out" of other systems like team, organization or company and to observe these systems externally. We enter into the universal system of systems. At this learning stage, a state of openness and connectedness is achieved with the "big picture". Stage IV therefore, also has a clearly transcendent dimension.

Example: Paul has become a senior project leader and realizes that even the sovereign use of a portfolio of project methodologies in the environment of uncertainty and suspense can result in poor project results. He asks the question "What entirely new approaches are needed to learn how to deal with uncertainty and suspense?" Through a synthesis of theoretical considerations, he builds hypotheses, creates experimental actions and adapts his new type of project management method accordingly.

Learning means change, so the question arises as to "What can we learn to adapt to (modified) environments?" Dan McAdams [4] identifies three areas of personality that are changeable to varying extents:

- The first area is that of largely stable personality factors – this hardly changes. It is part of an individual's "equipment", with its resources and limitations. In the Dilts Pyramid, this is the identity level.
- The second area is that of "characteristic adaptations", that is, individuals' fundamental convictions, values, cognitive patterns and beliefs, as well as behavioral strategies. This area changes both slow and abruptly. Changes at this level are often

experienced as intrusive experiences, where people suddenly see their life "with a fresh pair of eyes".

- The third area is that of the "life story". At this level, an individual integrates with all personality traits and experiences as a meaningful unity. This area is represented in the Dilts Pyramid by the vision, mission and belongingness levels.

Along an individual's lifeline, there are always times at which development starts in areas 2 and 3. At such times, where development is possible, they are characterized by information that is absorbed in a manner that fits the current "setting" or resource profile of an individual: The key, the information, matches either the lock or the profile, respectively. With such cases, large developmental leaps can be perceived.

As we can see, the level of identity and values and basic assumptions represent the main "bottleneck" for any development and higher stages of learning. The two views of the understanding of learning (Bateson's learning stages and McAdams' variable areas of personality) are well complemented by a third model: the four learning phases.

Figure 4.6 illustrates the four phases of learning and the related characteristics of support and delegation. The learning phases can be clarified using the following example: A small child is unaware that they cannot drive a car (Phase 1: unconscious incompetence). A child of 5–7 years will be aware for the first time that they cannot drive a car, and they would like to drive a car (Phase 2: conscious incompetence). A young adult, who has just passed their driving test, still experiences difficulties when driving, because driving has to be performed consciously (Phase 3: conscious competence). Only after 1–2 years does unconscious ease of driving emerge (Phase 4: unconscious competence). Learning phase 4 corresponds to intuition in automated processes.

According to our experience, the transition from Phase 1 to Phase 2 is a severe complexity driver. We all recognize such complexity drivers from puberty. But even in

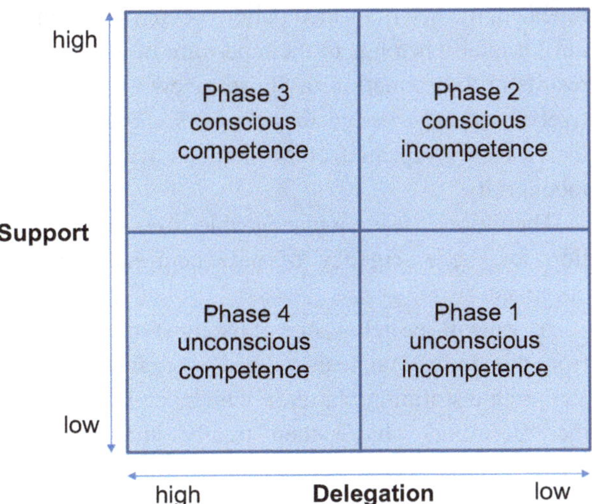

Fig. 4.6 Phases of learning

adulthood, corresponding situations still repeatedly occur where we are unaware of or vehemently block incompetence. The transition from learning stages I to II and the related dramatic changes in the meaning of personal values, basic assumptions and beliefs is one of the most drastic transitions from Phase 1 to Phase 2. In learning I there are many unconscious values, basic assumptions and beliefs, so a massive blockade to any change exists, since the need for change is not recognized at all.

In the learning Phase 1 it makes no sense for a leader to delegate and support. Such a need will not be recognized by those being "led". If the leader nevertheless delegates and supports, the blockade will be enhanced rather than degraded. The transition from Phase 1 to Phase 2 corresponds to that moment "it just clicked" or "you viewed it with a fresh pair of eyes": For example, a particular project manager was unaware for years that she had a significant weakness in her leadership competence: Only a dramatic project experience opened her eyes, and only after this situation was she able to reach Phase 2 and was ready to learn new things.

We refer to Grawe [5] and his remarkable work in Neuropsychiatry (see also Beck [6] for a unique pictorial representation of neural research results). He writes about this central learning mechanism from a neural point of view: "New coping strategies will be intensified and primed by the reduction of an existing incongruity (stress) …..Gerald Edelman (1987) in his 'theory of the selection of neuronal groups' explained the formation of new behavioral and experiential ways in the following manner:

In a situation of mental/neuronal instability in which no already established neuronal excitation pattern – he calls it neuronal group – exists that can degrade existing tension – stress or incongruity represent such tension – neuronal activity initially fluctuates back and forth between different possible states. Since none of these new states – i.e. neuronal constellations – are primed, these newly formed constellations will degrade quickly, until a constellation arises under current conditions, which reduces the tensions. This reduction of tension amplifies and primes the new neuronal excitation pattern. If this sequence is repeated, the newly formed pattern occurs more quickly and easily and it is better primed, until it finally belongs to the repertoire of an individual. What is important here, is that it results in the formation of entirely new neuronal patterns, which correspond to qualitatively new experiences or behaviors…The efficiency of the noradrenergic system is increased by repeated confrontations with challenging, but manageable situations of incongruity."[3]

Therefore, to learn, a manageable stress situation is required, because this is a precondition for the emergency of new neuronal structures. Without that stress, inertia is dominant.

As we will see below, one of the most important findings for transformation work arises from this. Individual, team-related or organizational transformation work is always associated with a storming phase, in which "confusion," i.e. stress, exists and then dissolved in the "norming" phase, then finally integrated into existing individual neural and

[3] Translation by the authors.

organizational patterns. Grawe also points out that transformation is only sustainable, if the related behavior patterns are repeated often. At this point, he emphasizes that existing neural patterns will not be (fully) deleted, but will be inhibited by the new neuronal patterns. This is also the reason why transformation can be destroyed, if existing systemic structures are not changed adequately: In such cases, the old behavior pattern fitting the old systemic structure will be recalled.

We would like to point out here the connection between learning and values. The Spiral Dynamics value-mem model [7] binds the ability of learning III and IV in particular, with yellow and turquoise memes. See also the appendix "Fundamentals Spiral Dynamics".

En Route to an Agile Organization, Part 3

Priesberg is sitting in a meeting with his colleagues, Horatio Wallace, Sue Snyder and Matt Broad, after having explained the directive in detail. Prior to this, the colleagues had been instructed on the topic of Collective Mind by a very experienced trainer. Priesberg hoped this would encourage their faster agreement. After about an hour, Priesberg asks his colleagues if they are willing to accept the content and the details and then subsequently asks them to develop their ideas on the flipchart. He is confident that he will be able to quickly reach a result where all three of his colleagues are in agreement and that he can then proceed to the next step: training.

Unfortunately, after only a short period of time, Priesberg's colleagues become embroiled in technicalities. Mr. Wallace appeared stressed: "So if you want to describe measurement procedures so rigorously, this means I have no more freedom at all. Why am I a biological-technical assistant then? I would like to have a say in when and how I measure something. And my department will see this in the same way." Sue Snyder interrupted him: "Well actually, I personally do not have enough detail. I would have extended the prescribed details to twenty pages. My department is renowned for carrying out measurements precisely. This is why so much weight is attached to our results." Mr. Broad leans back listening and contemplating minutes before expressing his opinion: "This damn Taylorism! Everything has to be meticulously regulated at every step. Then the result comes out right, doesn't it? Where is the overview? Who says that this step actually increases the quality of the test substances? I'm also struggling with my department on this topic, but that's another issue."

Priesberg notes that the discussion is no longer about content but about value propositions! It no longer makes sense to continue the meeting. He considers his colleagues ungrateful, selfish, and even arrogant. The training session does not seem to have had any affect at all. He closes the meeting with the words: "A uniform point of view is obviously not welcome here!" and intends to blow off steam with his colleague Ehrlich and to discuss project management generally and in specific detail. It's all pseudoscience!

Later on, when they meet up, Ehrlich listens patiently to Priesberg's tirade and then comments: "This is the difference between theory and practice. In theory, everything sounds simple. In practice, however, you need to be aware of the pressure points where the theory can be applied. And first and foremost, you need to start with yourself, i.e. with the acquisition of meta-competency."

Priesberg looks at him questioningly: "Not the Dilts Pyramid again?" Ehrlich continues, undeterred: "Of course! Because what you have delivered so far is an autocratic style of leadership. There is order and certain behavior is expected, which is finally controlled, without consideration of the individual sensitivities of those concerned. I call it 'learning zero'. It is the most primitive form of leadership, which does not meet the requirements of the modern project world in any way. I'm sorry, dear colleague, but I'm horrified. Did nothing stick from what we had discussed? Leading nowadays means being able to regulate a complex entity of elements."

4.3 Example of Leadership and Resonant Communication

There's nothing ill can dwell in such a temple:
If the ill spirit have so fair a house,
Good things will strive to dwell with 't.

William Shakespeare, The Tempest

The first step in learning III is the perception of one's own temperament and values, as well as the temperament and values of the individuals being led. The second step is to shape communication on this basis to achieve adaptive and situation orientated communication. Figure 4.7 shows two people, represented by their **personality flowers**. In this example, we assume that the person on the left is a team leader and the person on the right is a team member (of course, leadership can be reversed, a team member can guide their team leader; according to our understanding, leadership has little to do with stereotyped roles):

Conscious explicating of basic assumptions is part of meta-competency. In this book, we therefore define leadership based on the following assumption: Appreciation means to accept and develop the strengths of an individual and not to stress their weaknesses. For the above example, this results in the following leadership behavior:

The team leader regularly seeks contact with his team member. He is extroverted and accepts that his team member is introverted. While it seems simple at first glance, in practice it is not: The team leader might expect his team member to initiate contact in case of

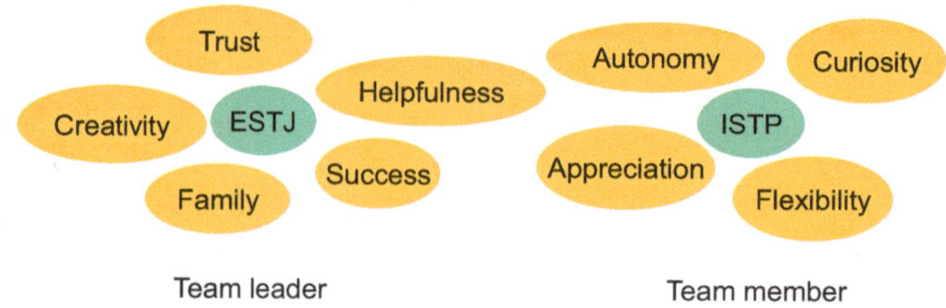

Fig. 4.7 Resonant communication

problems or questions. As an introvert may hesitate in making contact, this increases the danger of risks or other complexity drivers: The introverted team member will not challenge a decision or basic assumption on which work packages in a project are based, and therefore these decisions are not made in a conscious way in the team. In addition, in this constellation a certain type of common behavior often arises, which is very much underestimated: "One speaks and the other remains silent." In this case, it might be that regular meetings take place, but the extraverted team leader does not know what is going on in the team member's head, because as someone with a strong extroverted preference, he does not listen.

As the team leader is helpful, he offers a lot of support. However the team member has a high need for autonomy. Not infrequently the team member perceives support or the regular visits by the team leader as a violation of their own autonomy. This violation is then branded more firmly in their memory, the less the team member is able to sympathize with the need for this kind of support (for example, when they are in learning Phase 1 with respect to the required skills). – This may happen if the team leader wishes the team member to develop, but is unaware of the team members' learning phase.

For this reason, it is important, as a team leader, to be aware, whether support will be accepted and land on fertile ground: If autonomy is pronounced and the team member is unaware that they do not have the required skills, this will drive social complexity between both customer and team member, as well as between team leader and team member. It will stir resistance, and in the worst case scenario, the team member will build an aversion to the team leader and perhaps even towards the customer.

We also emphasize self-leadership: If a team leader is aware that they are being helpful, they will regulate their behavior, not only with respect to the team member, but will also adapt their behavior with respect to their own needs. The team leader will therefore take into account the learning phase of those being led and offer as much help as is appropriate for their own well-being and to ward off any exploitation attempts.

If we look at the MBTI values of the two individuals above, we can see that the team leader's temperament has a J-preference: For example, those with a J-preference tend to start performing a task long before the required end date and normally finish well before it. This is in marked contrast to the behavior of the team member with a P-preference. This team member will estimate the effort required for the task and begin the task so that it is completed close to the required deadline. A competent team leader will be aware of the difference in strategy for task processing: Although insisting on meeting the deadline, they will also accept a last minute presentation of results. If instead, they (incompetently) keep pestering their team member for the status of the task, the team member will feel hurt with respect to his temperament and his "autonomy" value.

Furthermore, the team leader will ensure that they assign their team member new and interesting tasks (curiosity and P-preference) and praise team member for their good work (appreciation).

At this point we would like to mention an observation that we have made in many communication situations and trainings: Values like "to attach" to a suitable temperament profile. In the case of the team member, "flexibility" and "autonomy" fit very well with their P-preference. But we have to be careful: No rules can be derived, according to which certain values or motives will always occur in connection with certain temperaments. This is understandable, as motives and values represent categories of motivational schemes and are concretized by them. For example, the motivational scheme of the value "creativity" can differ significantly for different individuals. However, the motives and values help to set a focus in the large number of motivational schemes and to induce communication in the team in a structured way, about values and motives.

The communication that is aligned with the personality of an individual is the central pillar of resonant communications. In the Collective Mind Method [1] we developed the "transformer-model" for communication that, in addition to MBTI (incl. the meta-programs), integrates the four-ears model, the transaction model and the NLP meta model of language. We are now extending the transformer-model to the model of motives and values, and to the model of basic assumptions.

This results in enormous power for modeling the variety of communication and therefore for the regulation of communicative complexity. In our experience, the extended transformer-model is a very good comprehensive model for "normal communication" between individuals. We speak of "normal communication", because the transformer-model cannot be applied in situations where mental disorders arise, as it does not contain appropriate models.

En Route to an Agile Organization, Part 4

Ultimately, Priesberg intends to take Ehrlich's remarks seriously. Ehrlich had stated that 'The values and basic assumptions of the team members and, subsequently, of everyone involved, must be consistent with those values and basic assumptions provided by the directive' – just after Priesberg had marched off indignantly. The idea sounds good, but he is puzzling as to how to carry it out. Priesberg checks his e-mails and finds the following message from Mr. Freschi, marked as strictly confidential:

"Dear Mr. Priesberg, I am a appalled, to say the least. With the first draft of the directive, you have caused more unrest than any useful effect. I had to urge three of my department heads to stay in line, because they were about to rebel, based on the feedback they received on your last meeting. In the end though, something good will come of all this: the management will set up a governance unit, to issue directives in research, similar to a legislative force. It will function as a setting parameter and decide on control parameters. I continue to assume that you too consider the project to be complex. This will hopefully enable us to establish an order parameter, namely, the similarity of measurement results for the same test substances. This means that for all employees involved, there will be no escaping the directive, and of course this will increase pressure on you considerably, but I am confident you can do it. Yours, Herbert Freschi."

Priesberg gulped when he finished reading this. He had rarely experienced so much reproach along with the simultaneous expression of trust, yet for the first time realizes how Freschi creates a basis for self-organization. He even seems to know what he is talking about, and expects the same from Priesberg. But, he feels there is a lack of concrete elaboration, which he is unable to articulate further. He decides to visit his colleague Ehrlich immediately.

On the way to Ehrlich's office, Priesberg bumps into Mr. Broad. "Well, well, is Mr. Controller on the road checking and measuring everyone's steps? I personally have come to a final opinion on the new directive – I also believe that I am not on my own: It seems freedom of decision is no longer desired," he says angrily as he passes by.

Ehrlich addresses the reaction of Priesberg in the meeting with the three key stakeholders: "By your remark 'unity is not desired here', combined with borrowed power from Freschi, you created a negative resonant intervention and set a control parameter, through which value-destroying complexity comes into being: 'freedom of decision is not desired here', is the pattern that has developed as a result. You have therefore given the directive negative connotation and you have to subsequently expect maximum resistance from now. Whenever you want to promote the topic, all those involved will resist. There will be discussions, alliances, and blockades against the directive. A complex structure will emerge that works against the meaning and purpose of the directive."

"How can this have happened," asked Priesberg in astonishment. "It's quite simple: a research company must give its employees certain freedoms. These are accepted as their values and basic assumptions. And these include a certain skepticism of regulations and directives. This is a very natural reaction, and value-destroying complexity emerges," concludes Ehrlich, feeling enthusiastic, because he observes Priesberg's willingness to learn. Priesberg is still feeling around in the dark, but has realized that his current position will not lead to a solution.

"How do I get out of this dilemma?" Priesberg groans. "Freschi has already secured the front line with the governance unit, he has one eye on complexity and self-organization. However, he has left us with the details: the directive must presumably be completely re-established," states Ehrlich.

Priesberg looks horrified, he feels as if he can neither move forward nor backward. Ehrlich pats his colleague on the shoulder. "Do not take it so tragically, we will talk about how we can structure the directive as a whole."

4.4 Example of Leadership and Value-Destroying Complexity

...Tell them that God bids us do good for evil:
And thus I clothe my naked villainy
With odd old ends stol'n forth of Holy Writ,
And seem a saint, when most I play the devil. ...

William Shakespeare, King Richard III

In this section we want to look "below" the surface of leadership by using an example. We will do this by considering the systemic implications of the personality of the leader on their leadership style. Only the previously mentioned models and theories will be applied.

When applying theories and models, it is almost always necessary to limit potential variety. We have shown that the theories and models proposed in this book have a variety far beyond any practicability when exploited completely. But experience helps: In many cases it is sufficient to model leadership style by using three dimensions.

These three dimensions are derived from the main leadership styles discussed above, by assigning each dimension a continuum of characteristics. Figure 4.8 describes these three dimensions. The dimension "degree of alignment" is defined by the manifestations "leading by actions", "leading by objectives" and "leading by vision & mission". The dimension "degree of attention" is described by the two poles of "transaction oriented: performance by incentives" and "value oriented: performance by shared values". The dimension "degree of participation" is characterized by "autocratic: leader marshals", "participative: information sharing, leader decides" and "democratic: team decides". These three dimensions represent with experience, preferred specific views of leadership on the Dilts Pyramid levels: Vision & mission, values and basic assumptions and behavior.

Let us turn to our example: A leader feels that their style of leadership is not understood by their team. Not infrequently, this has resulted in massive clashes, partly in conversations with team members, but also partly in team meetings. A team coach is consulted who will

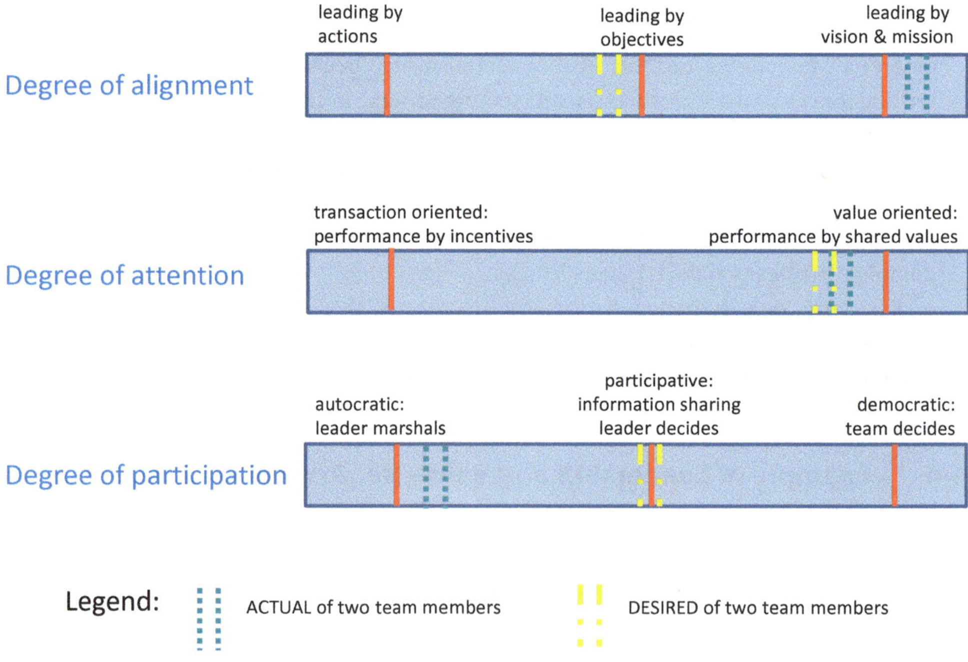

Fig. 4.8 Dimensions of leadership

attend some of the team meetings. After a first team meeting, the team coach generates a hypothesis about possible root causes. To test the hypothesis, he suggests holding a workshop so that the team members can experience MBTI and in addition the leadership style will be analyzed by private questioning of each team member (of course, the approval of the leader must be given in advance). Based on the above model the team members indicate both perceived leadership (ACTUAL) and the type of leadership they desire (DESIRED).

Figure 4.8 considers only two team members for the sake of clarity. After each team member has entered their assessment in the graph, the result is presented to the team members and the leader. As can be seen in the figure, team members perceive a visionary leadership style, but desire a leadership style more oriented to specific objectives with a clear tendency to individual tasks. With respect to degree of attention, there is almost no change between ACTUAL and DESIRED, this means a value oriented leadership style is perceived, which requires almost no change in this dimension. With regard to degree of participation, in this example the leader is shocked: An autocratic leadership style is perceived, which needs to be replaced by a participatory style. This is despite the fact that the leadership style was one with a high level of employee appreciation.

How can this be?

What is striking in this example from the consulting practice is the high level of autocracy with simultaneous value orientation. This would seem to be a contradiction. Such a contradiction can only be resolved if parameters are found that will help resolve this. As we will see in the section "Transformation Management", these parameters can be found at the micro-tier and/or the macro-tier and could suggest necessary interventions: When observing communication patterns between the leader and team members, it becomes clear that the perceived autocratic leadership style has its root cause in the systemic constellation of the personalities involved. The temperament of the leader shows a clear ENFP preference, which means extraverted, strong visionary, friendly and relationship-oriented and always open to new ideas.

The friendly and relationship-oriented communication style of the leader results in very appreciative behavior, in addition to socializing in private. However … there is an additional aspect. The team members are more practical, i.e. with respect to MBTI they are ST or SF-oriented. They can do little or nothing at all with the visionary, abstract communication of their team leader. In a relaxed private environment (breaks, evening meetings) these differences do not occur.

In a professional context, very often situations arise where the leader loses their temper, because the leader realizes that they are not getting through to their team with their remarks: They give the appearance of carrying out an autocratic leadership style, although that was never their intention. In such cases, they very often have no other choice than to impose their ideas using dictatorial statements. However if the team members have an intuitive temperament orientation (NF or NT) they would most likely perceive the leadership style as democratic.

From this, it can be seen that the systemic coupling of different personalities leads to a macro-structure of leadership, which is neither desirable for the team leader, nor for the team; and is even diametrically opposed to the personality preference of the team leader. We have referred to such a macro-structure above with the phrase: "the whole is less than the sum of its parts".

From this, a pattern can be deduced, from which complexity emerges: When there is a lack of ability or energy to align one's own communication to the situation – we speak of a lack of internal resources – which creates a value-destroying complexity in the overall system. This results in a dysfunctional strong coupling between the control parameters, emerging in a value-destroying situation: In our example, the visionary "degree of alignment" of the team leader creates a decoupling of team members from communication and this decoupling is perceived by the team leader as their own "powerlessness". In response to this, the "degree of participation" moves in the direction of an autocratic leadership style.

In summary, a much more general statement can be derived. We have already stated this in the chapter "Social Techniques and Complexity", as well as in the appendix "Fundamentals Theory and Practice": Without appropriate theories and models, such an example remains only "in the depths of the ocean". In our example, even a single model would be completely inadequate; it is necessary to connect different theories and models to obtain reliable statements. With the number of models given in this book, we already have a "myriad" of connections, so it is necessary to create a preselection based on practical experience and intuition, i.e. to form hypotheses. Without theories and models, any intervention in the system "leadership and team" of our example would be at best ineffective, and in the worst case harmful, supporting value-destroying complexity.

4.5 Neuroleadership

We know each other's faces; for our hearts
He knows no more of mine than I of yours,
Or I of his, my lord, than you mine. …

William Shakespeare, King Richard III

En Route to an Agile Organization, Part 5

At his next meeting with Ehrlich, Priesberg becomes irritated: "Before we think about the systemic environment, as you are fond of calling it," he smirks, "Don't you think we should talk about the facts first. Here's what I find strange: In Mr. Wallace's department, samples are stored in a professional manner, but measured at different times, relative to the time of receipt of the sample. Mr. Wallace has also developed his own measuring method, of which he is particularly proud. In Ms. Snyder's department, the measurement apparatus does not appear to be calibrated according to DIN,[4] but according to another norm, which is not so common. And in Matt Broad's department they do

[4]DIN is the German Institute for Standardization.

not really seem to care about thorough documentation and analysis, for them monthly traffic light labeling reports for department management is sufficient. The traffic lights are usually on green, because this is more acceptable: Critical dots seem to disappear from the field of view and lead to blind spots. You could say that traffic lights are management parameters that lead to value-destroying complexity. The measurement results and detailed reports are then so different that a comparison is extremely difficult and yet this does not seem to disturb anyone. And so these final test results from the three departments can make further decisions very difficult or create mistakes which will cost a lot of money later on." Priesberg has now calmed down a little from his initial state.

Ehrlich suggests: "Does it not make more sense then to have one single department responsible for measurement? Then these problems would be immediately eliminated." Priesberg retorts: "It's not that simple. Each department specializes in specific active ingredients and is directly assigned to respective research departments, as well as being very closely linked to their sample logistics. A new policy would appear to be of more help than reorganizing the departments."

Ehrlich continues: "So the problem cannot be further simplified. Next, we need to consider how the directive fits into the work environment of the employees, each working in departments with different departmental cultures. Finally, the directive is supposed to guide, and this is only possible if it generates motivation. So now we have to deal with motivation and motives. Otherwise, any effort is in vain, because personal motives very much determine individual ways of acting."

Motives and values are the key factors in the Dilts Pyramid "central block". They are quite significant in determining transparency and flexibility in the mindset of an individual or organization. They are probably the most important control parameters in the system "man", as well as in the system "organization". Only with sufficient "permeability" can the actions of individuals align to a vision.

Neuroleadership is an approach whereby this insight is integrated into a holistic approach.

With regard to our understanding of Neuroleadership, we orient ourselves to Peters and Ghadiri's definition [8]. Peters and Ghadiri define it as follows: "Neuroleadership is the designing of a working environment for employees, in which the leader deals with organizational and personnel management measures to generate a "brain-compatible" environment according to employees four basic needs."[5]

This understanding of Neuroleadership is based heavily on fundamental research in neuroscience and psychology, which has been largely influenced by Grawe [5] (see appendix "Fundamentals Consistency Theory"). In Grawe's Consistency Theory, there are four basic human needs that define our actions. If they are not serviced, this results in performance reduction or even lead to many of the mental disorders that are known to us.

[5] Translation by the authors.

We strive to continuously align our basic needs, i.e. to achieve consistency, with our experience and our actions by so-called motivational schemes. Motivational schemes are specific behaviors that we have acquired through experience. They can be differentiated by approach and avoidance schemes. Through ongoing experience they can be enhanced or inhibited by other experiences.

It is important to note that our memory does not delete experiences we have made in the past. Experiences of any kind are therefore not deleted by new experiences, but only inhibited. Depending on the context, old experiences are activated to a greater or lesser degree. Inconsistencies between needs and experiences, depending on their strengths and nature, can result in mental disorders. For all four basic needs it is recognized that the experience of consistency is accompanied by the outpouring of neuropeptides in the brain. Closely related to this is the central neural mechanism of learning that we have cited in the section "Learning and Meta-Competency".

Figure 4.9 illustrates the four basic needs to which we related the 16 motives outlined by Reiss (see also appendix "Fundamentals Reiss Motive Profile") and the four levels of the Spiral Dynamics culture and awareness model (we refer to appendix "Fundamentals Spiral Dynamics"). Until now no accessible scientific validation of the 16 motives has been published, including for the current assignment. However, in the consulting practice of one of the authors, the 16 motives have been proven in conjunction with values. We consider motives (and values) as categories that bundle[6] similar motivational schemes together. In the brain, rewarding systems will be activated according to the satisfaction of basic needs and their associated motivational schemes (simplified: motives and values). Rewarding systems are complex neural processes with strong (happy) feelings, accompanied by neurotransmitters and neuromodulators like serotonin and dopamine [5, 9, 6] among others.

Motivational avoidance scheme examples:

- I have so much to do and therefore I have no time for this project.
- I don't like to be just a little cog in a big wheel.
- I don't have the necessary know how and therefore will not pass any judgement.
- I don't like large meetings with many participants that have no value anyway.
- I don't know these people – I'm suspicious of them.

Motivational approach scheme examples:

- I like to work in peace and quiet.
- I prefer to work on topics of interest to me.
- I like to decide where I want to go.

[6] This is again a form of abstraction, which we use to train intuition, because with bundling, we bundle schemes with similar patterns.

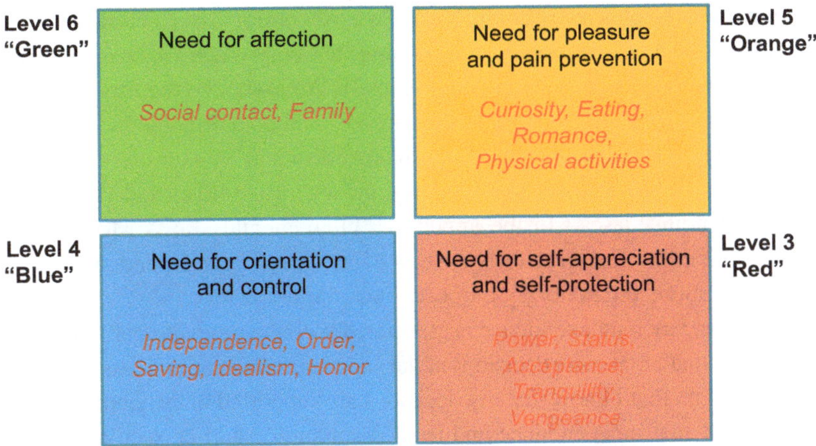

Fig. 4.9 Basic needs, Reiss motives and level 3–6 of the Spiral Dynamics model

- I like to work with people I know and appreciate.
- I love to be able to include my competencies.

As a small exercise for the reader, we recommend assigning these motivational schemes to motives. Similarly to the temperament profile, the Reiss Profile is based on a self-assessment questionnaire. The questions contained in the questionnaire are stereotypical examples of motivational schemes. Let us take a closer look at the four basic needs (see also appendix "Fundamentals Consistency Theory"):

- The need for affection represents our need for human contact and closeness. The motivational schemes associated with the need for affection are acquired to a large extent in the first years of our life. We assign the motives social contact and family to this basic need. In a professional environment the need for affection is also linked to team orientation ability, as well as to the ability to establish trust and consensus in professional collaborations and cooperation.
- The need for orientation and control is very strongly linked to the interaction of our self with reality and our perception of reality. Our experience teaches us whether we will achieve our goals and how we will achieve them. By making positive or negative control experiences, goal-oriented activities and control are inseparably linked to each other. The need for orientation and control is also linked to our sphere of influence and of safety. The motivational scheme category "saving", also simply referred to as the motive or value "saving", could on the one hand be associated with the desire for safety. On the other hand, "saving" can result from the desire for independence, which is made possible by "saving": People for whom "saving" is important, have learned in the course of their lives that "saving" helps them to control their resources and thus gain freedom for other actions; therefore, control is exercised in an indirect manner. Closely linked to the need for

orientation and control is also our ability to learn. According to the central neuronal learning mechanism, stress is important for learning, but only when we have it under control (see section "Learning and Meta-Competency"). We associate the motives of independence, order, saving, idealism, honor and purpose- and goal orientation with the need for orientation and control. Using the motive independence as an example, we would like to point out that the bundling of motivational schemes into motives and values contributes to the risk of noun biases.[7] In the appendix "Fundamentals Reiss Motive Profile" we mentioned that in the literature on the Reiss Motive Profile, independence was replaced by team orientation, and honor by purpose- and goal orientation.

For example, when independence is associated with the motivational schemes "for me, it is important, that I make my decisions alone" and team orientation is associated with "for me, it is important that I know that my decisions are supported in the team", then it is quite understandable, that independence and team orientation could be seen as two poles of a motivational scheme. If team orientation is linked to the motivational schema "I like to work in a team, because I like to have people around me", it would be appropriate to assign team orientation to the basic need for affection. In this example, we can clearly identify the limitations and risks of model building based on a focus on large abstractions.

- Among the four basic needs, the need for self-appreciation and self-protection is regarded as a specific human need, which does not occur in animals. This is because this need presupposes a consciousness of ourselves as individuals and the ability to think reflexively.

  All humans have the need for self-appreciation as convergence schemes, even if self-appreciation is abandoned in the context of satisfying others' needs: For example, the need for self-appreciation is still present when the need for affection necessitates self-humiliation.

  If important characteristics of an individual's self-image are challenged by the environment, self-humiliating statements will be "forgotten" for the purpose of self-protection. This acts as an avoidance scheme.

  Self-appreciation and self-protection are closely linked to cognitive biases [5]: "Actually, those who are psychologically healthy are those, who have a biased perception of reality with respect to themselves and not those with poorer mental health."[8]

  Even unrealistic optimism with regard to oneself has its roots here [5]: "The probability of having a serious accident is considered to be much lower than is actually the case in reality."

  We assign the Reiss motives of power, status, acceptance, tranquility and vengeance/fighting to this need.
- The need for pleasure and pain prevention is an expression that we are striving for pleasant states and avoiding unpleasant ones. We are constantly evaluating our experiences with regard to the qualities of "good-bad". What we feel as pleasant or

[7] We speak of a noun bias when a process and/or one or more behaviors are grouped together in a term, thus losing the underlying process or behaviors.

[8] Translation by the authors.

unpleasant is learned to a not insignificant extent. Tastes and odors, among other things, belong to learned preferences: Bitter (i.e. beer, coffee) and hot (i.e. chilli) foods, for example, become soft foods, that we experience with pleasure via socialization. We assign the Reiss motives of curiosity, eating, romance (sexuality), and physical activity to the need for pleasure and pain avoidance.

Peters and Ghadiri [8] propose carrying out a questionnaire that asks about fulfillment of current needs in the workplace or in projects, and then compares it to the employees' desired fulfillments. In Fig. 4.10 the ascertained consistency profile illustrates the diamond layout that is known for the actual situation (red) and the target situation (blue).

The questionnaire by Peters and Ghadiri [8] asks about the four basic needs by using typical motivational schemes. For details of the questionnaire, we refer to Peters and Ghadiri [8].

The goal of Neuroleadership is to make actual and desired fulfillment coextensive as far as possible. To achieve this, in principle, there are two conceivable ways:

- The work environment of the employee is designed so that it (better) fits their needs. In the example in Fig. 4.10, the employee desires less team integration and more peace and quiet and a much more challenging task that will take into account their curiosity.
- Transformation work is started to review the motivational schemes for the need for lesser team integration and more peace and quiet. This transformation work gets to the bottom of the avoidance and approach schemes and attempts to effect an adequate transformation, but always with due regard to appreciation of the employees' personality. If the leader does the job themselves, they should be qualified as a coach. It should be noted that transformation work by the coach must not be mixed with therapeutic work, because this requires a quite different qualification.

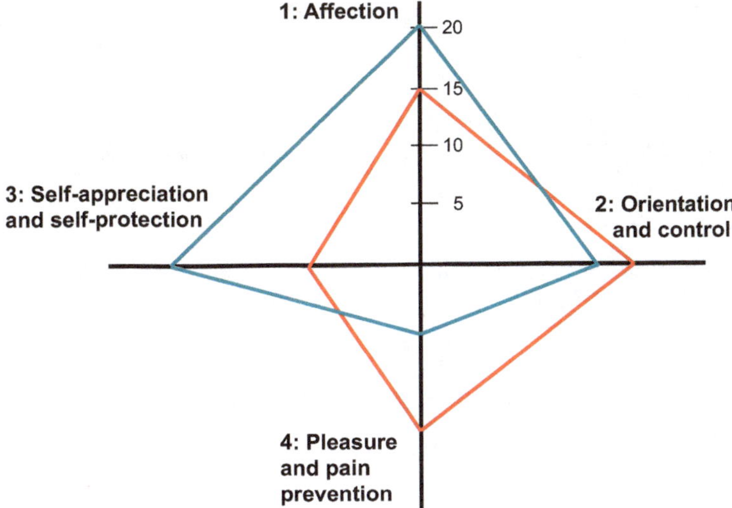

Fig. 4.10 Consistency profile of the Neuroleadership model

Neuroleadership requires the leader to have a very high level of personal skills, because the leader acts less as a typical manager or consultant, and more as a coach. We will return to this distinction in the section "Transformation Management". In terms of self-organization, coaching means adjusting the setting and control parameters in such a way so that self-organization can emerge. The setting parameters are mainly set by adapting the temporal and spatial components of the work environment (for example, introducing periods of quiet working, team meeting rituals or "lockable" rooms).

By adjusting the control parameter "personal appreciation of individual motives and values", the foundations will be set for forming the flow state [10]. This also means that the transformation limits are set according to employee personalities.

As we have already seen, values can also act as order parameters. This is especially the case when they act as cultural elements. Duchmann and Töpfer [11] reformulated the four basic needs as four ranges of basic values of an organization. The need for affection manifests as consensus values, the need for orientation and control is expressed as bureaucracy values. The need for self-appreciation and self-protection is reflected in competition values, and the basic need for pleasure and pain prevention is reflected in entrepreneur values (Fig. 4.11).

A direct connection to the value-memes (v-memes) of Spiral Dynamics [7] are given here (see appendix "Fundamentals Spiral Dynamics"). The term "meme" is based on the concept of "genes". Just as genes are the basic building blocks of our DNA, so memes are thought snippets, the basic building blocks of our thoughts [12].

For this reason, in the previous figures the color codes of Spiral Dynamics were used for the different v-memes (red, blue, orange, green). It should also be mentioned that there are different cultural models using this classification with slightly different content and

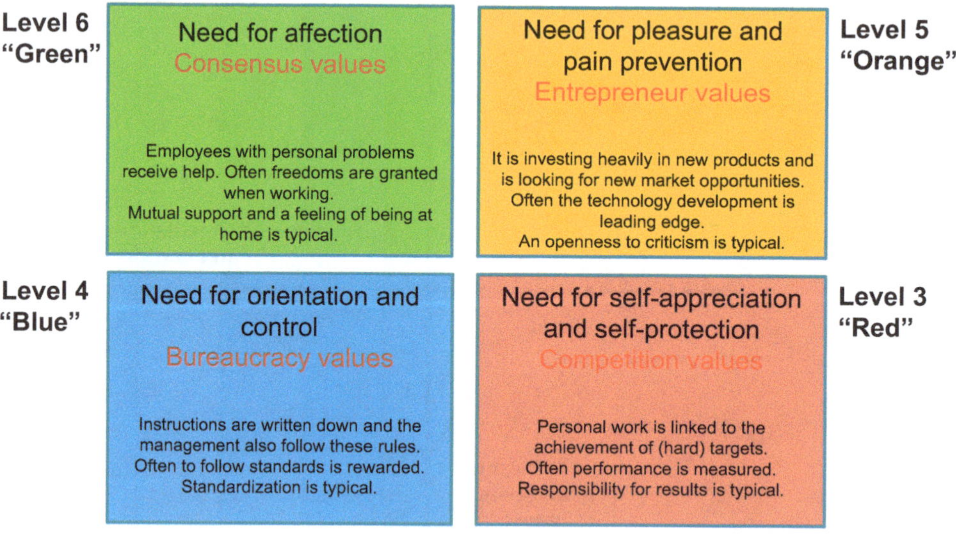

Fig. 4.11 Neuroleadership, company culture and level in the Spiral Dynamics model

slightly different classification names [13, 14]. This is an example of where for some aspects, a number of models exists, expressing similar content, but rarely showing advancement. As mentioned elsewhere, we urge the use of a diversity of models where the driving force is the gain of insight, not licensing politics.

4.6 Team Leadership, Intuition and Biases

Hastings:
Good morrow, Catesby; you are early stirring.
What news, what news in this our tott'ring state?
Catesby:
It is a reeling world indeed, my lord,
And I believe will never stand upright
Till Richard wear the garland of the realm.

William Shakespeare, King Richard III

En Route to an Agile Organization, Part 6

After extensive consultations with Ehrlich, Priesberg met up with his three colleagues, Mr. Wallace, Ms. Snyder and Mr. Broad again. He promises them that he will take their individual situations into account when drafting the directive. The mood is subsequently lighter and Priesberg conducts a thorough interview with each of his three colleagues. This results in a directive of twenty pages instead of fifteen. Priesberg now believes that all motives and needs have been taken into consideration and feels very confident …

Three days later, Priesberg wishes to discuss the directive with his colleagues. He invites Ehrlich to the meeting, which also includes Dan Leibowitz, a representative from internal revision and an old acquaintance. Once the meeting had been opened, the three key stakeholders began immediately: "So, I see that my details are now represented," commented Ms. Snyder. "And my competence as a biological-technical assistant has also been adequately taken into account," adds Mr. Wallace. "Well, I am not really sure" adds Mr. Broad, "as so many paragraphs are formulated in the subjunctive, I can just simply omit them." Mr. Leibowitz takes the floor: "At this rate, we will not get any further with the wording of this directive, it is ambiguous and allows too many freedoms. Mr. Priesberg, what is the reason you want to formulate the directive so extensively?"

Priesberg is surprised: "As a biotechnologist, I would feel bad if it were not worked out exactly. I see no other way than formulating the directive in this way." The key stakeholders nod eagerly.

Mr. Leibowitz, who is very well versed in complexity, replies using specialist jargon: "We should imagine the future project results, in fact a negative result. By using conscious reframing,[9] the missing control parameters can be seen. A setting parameter has already been given through the governing unit." Except for Ehrlich and perhaps also Priesberg, all the other participants seem slightly irritated.

Priesberg feels that all this means a great deal of change to his previous considerations. For him, his version of the directive is still an 'all-rounder'. However, it would seem to carry the risk of not creating a stable Collective Mind. Ehrlich murmurs quietly, it seems to be a case of telepathy: "This directive, including training, will come to nothing. Priesberg is obviously biased and guided by his biotechnology glasses. In addition, as a fact-oriented type (MBTI dimension 'S'), he tends to pursue what is known. We need to systematically eliminate these distortions, and at the same time carve out the system control parameters under construction, in order to increase the quality of the measurement procedures."

Biases are a great source of value-destroying complexity. They occur in individuals, and are therefore also present during teamwork. As we have seen in the section "Interplay of Intuition and Rationality", Kahneman and Tversky described relationships, and insights into our behavior, which are not accessible through our perception alone using their theory on the psychology of decisions (Prospect Theory). This again shows that the insight of "we do not know what we do not know" is also valid here, and that this lack of knowledge is crucial in creating complexity that does not create value, but actually destroys value and therefore contributes to our uncertainty. Applied to biases, this means: Very often we do not even know that we make decisions based on biases and this unawareness creates effects in projects, for which we have no explanation. As we have shown earlier, biases occur as individual errors of reasoning. In a team, these biases usually occur unconsciously and vary from person to person. The result is a value-destroying complexity, because different individual biases interact in a non-predictable way.

We can illustrate this by adopting our explanations from [15] and [16]. With Systems 1 and System 2, Kahneman provides a tool that allows individuals to detect biases in decision processes, from which they can derive appropriate countermeasures. This raises the question, whether and how it is possible to apply these instruments to decision processes in groups, and finally in projects.

In other words, the question arises as to whether team-oriented instruments exist, based on "Thinking, Fast and Slow" [17], to help minimize unwanted biases and to exploit desired biases in projects.

From this fundamental question, we have derived the following questions:

- How can projects be structured so that individual and collective biases can be detected and reduced?
- Is it possible to increase value-creating complexity while minimizing value-destroying complexity and thus reduce uncertainty?

[9] Reframing means a reinterpretation or a "new contextualization", with the aim of avoiding biases and activating mental resources.

- Can the processes in a project be organized so that individual and collective biases are minimized?
- Is it possible to weigh the positive properties of System 1 and System 2 according to the type of project and project phases, respectively, so that they can deliver value in a project?

Let Us Turn to the First Question How can projects be structured so that individual and collective biases can be detected and reduced?

It is quite striking and surprising that in his book, Kahneman does not connect any dots to the psychology of personality, although the central terms of intuition and rationality seem to be very similar to the S-N temperament dimensions ("intuition" (N) and "sensing" (S)) of the MBTI (however Kahneman strongly advises the use of systematic criteria catalogs for the assessment of individuals to escape the mechanisms of biases).

In MBTI, the S-N dimension is characterized by pole sensing (S) and intuition (N): The S-pole represents a detailed and fact-based personality preference; the N-pole represents a holistic, big picture-oriented personality preference. See also the appendix "Fundamentals MBTI".

We have formulated a hypothesis that a personality preference for intuition N means a strong preference for System 1, and a preference for sensing S means a strong preference for System 2. We encourage verification of this by appropriate quantitative studies. Qualitative confirmation of this hypothesis stems from observations we have made in our professional or consulting practices, respectively.

Assuming that this hypothesis is correct, it is useful to build teams so that Systems 1 and 2 can work to their best effect there, where they can play to their strengths. In an innovative, complex project, it is useful at the beginning, when the Goal-tier is being searched for in the solution space (see Fig. 4.12), giving more weight to team members with a

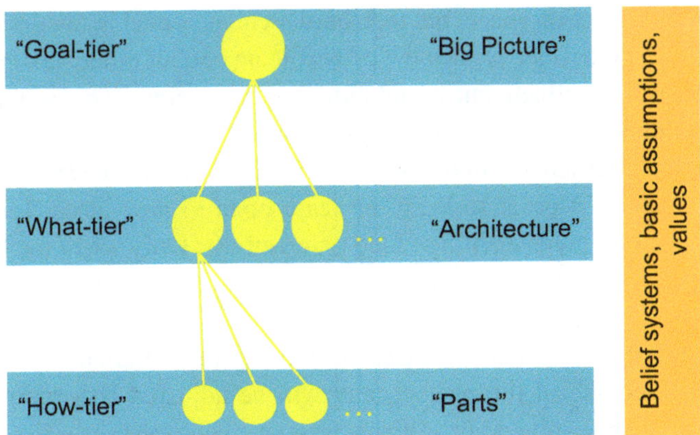

Fig. 4.12 Goal-, What- and How-tiers for linking together the mental models of project team members

preference for System 1 in the project team and at the same time it is being ensured that the openness associated with this result does not enter into a bias too early.

It is one of the key tasks of the project leader to take care of this by suitable interventions (for example, by decorrelation, Walt Disney strategy, etc.). The interventions which open the solution space are supported by individuals with the MBTI preferences NP (for intuition and perceiving) in the team. People with the NP-preference adapt very quickly to situations and love to bring in new ideas and highlight them from all angles.

The over-emphasis on individuals with a preference for System 1 (N-preference) at an early stage of the project should be "corrected" during the early stages by individuals with a preference for System 2 (S-preference): After having found a viable solution in the solution space, it is important to initiate and support target orientation. At this stage of implementation, it is advisable to over-emphasize the share of people having a higher preference for System 2 (S-preference).

Let Us Now Turn to the Second Question Is it possible to increase value-creating complexity while minimizing value-destroying complexity and thus reduce uncertainty?

Earlier, we showed that self-organization is the preferred tool to stimulate the development of value-creating complexity. For this purpose it is necessary to shape "Principle 3 – creation of sense and meaning" into an order parameter.

In order to achieve a connection of the mental models of team members according to the Collective Mind model [1], it is necessary to map the desired solution of the project by the three tiers: "goal", "what", and "how". Figure 4.12 shows the Goal-tier containing the "Big Picture" of the project. The middle What-tier contains the architecture of the target hierarchy. The How-tier contains the design of the single steps that specify the target fields of the target architecture. The Goal-tier is the tier of the big picture and acts as an anchor for System 1, to which the other two tiers align.

The Goal-tier (as are the other two tiers) is determined after an exploratory search phase in the possible solution space, but then hardly changes during the project (in contrast to the How-tier). The Goal-tier primes the project team and all other stakeholders towards a common goal ensuring effectiveness and reducing uncertainty. The establishment of the Goal-tier as an anchor may only then take place, when the solution space has been sufficiently explored for alternatives. In this search phase, it is always necessary to apply measures for decorrelation in order to keep the team open to new ideas. John Cleese [18] explains the open mode in his exciting video on creativity. Only when the Goal-tier exists without doubt, as a "Big Picture", can the mental opening be reduced and turned into closed mode.

The content of the How-tier (the tier containing the detailed solution steps) can change during the course of the project, and will be adjusted according to the setting conditions and insights gained during the project. This ensures efficiency. With its details and facts of the target achievement, the How-tier ensures a minimization of biases at the operational level, because this tier refers to System 2 (rationality).

The elaboration of targets using the SMART model belongs only to the How-tier: If the "big picture" of the Goal-tier is replaced by a too detailed elaboration, the Goal-tier loses its central role as an anchor.

The specialty of the Collective Mind approach is the affinity of specific tiers to different personalities. Thus, the Goal-tier, which is dominated by big picture elements, is preferred by intuitive temperaments (N-preference, System 1) who quickly abstract and generalize from a concrete context with their disposition for intuition. In contrast, the How-tier is preferred by sensing-oriented individuals (S-preference, System 2). They have a high demand for detailed analytical and logical structuring of a problem and prefer to work in a concrete context.

On the What-tier, both preferences are interconnected, and they are able to communicate. By the introduction and application of the Collective Mind three-tier model, the cross-linking of mental models in the team is enabled and this generates value-creating complexity. Value-destroying complexity, such as biases as a result of focusing on "wrong" solutions, or communicative misunderstandings in a project team, will not occur or will be almost completely reduced.

An important point to make when designing the target hierarchy, is that all target fields in the What-tier are reflected in concise form in the Goal-tier and that the objectives of the How-tier are clearly assigned to a target field of the What-tier. In this way, a network-tree for the target hierarchy with roots on the Goal-tier will be created. During elaboration of the target hierarchy in the team, beliefs, rules of thumb, biases, basic assumptions and values in team communication will become more and more visible.

These need to be disclosed, because if this does not happen, the target hierarchy will probably be accepted only in a superficial manner, and in addition there will be no emotional and rational sustainable basis. Unrevealed individual values and basic assumptions hamper true integration of the target hierarchy in the minds of team members. This will soon become evident in the project and will consequently result in value-destroying complexity. It is therefore advisable to document all values and basic assumptions related to the target hierarchy.

Let Us Now Turn to the Third Question Can the processes in a project be organized so that individual and collective biases are minimized?

We propose the following strategy, generalizing Kahneman's proposal:

- Step 1: Apply System 2 and collect information (from inside view and outside view, decorrelate and reduce early priming by using facts).
- Step 2: Apply System 1, i.e. creativity and intuition, to create something new based on information.
- Step 3: Apply System 2 to review and test the results of creativity and intuition.
- Step 4: Apply both Systems 1 and 2 and perform a rational and emotional review of the effects of the results as a sort of holistic view.

This strategy corresponds the Z-Model process, which we used already in the Collective Mind Method [1], and is extended according to the bias dimension. Figure 4.13 illustrates the Z-Model from the perspective of the proposed strategy:

The first step in this strategy is the selective gathering of information. This is information from an inside view, but also in particular, from an outside view: If a company starts a specific type of project for the first time, which is not in the current portfolio, it is advisable to acquire benchmarking information (statements on costs, efforts, risks, etc.), from external projects of this project type. The aim of this strategy step is to provide System 2 with information, so that System 1 cannot gain the upper hand.

In strategy step 2, System 1 prevails. Based on the information from the previous step 1, System 1 creates new ideas and concepts with respect to the target.

In strategy step 3, the results of the creative phase are reviewed by System 2 and discarded if necessary.

In the final strategy step 4, the identified possible alternative solutions are tested for their effects in the stakeholder environment. This corresponds to a systemic view of the project.

This four-step strategy can be repeated several times until one or more suitable solutions are found.

Kahneman's strategy process for minimizing biases, includes the main elements of the Walt Disney creativity technique. A look back to the section "Regulation of Complexity Through the Formation of Models and Intuition" shows that the first three steps of the Z-Model correspond to Einstein's technique for creative problem solving by intuition. In Einstein's model, the "intuitive perception" step in Fig. 4.13 corresponds to two steps, and the last step here, which is oriented to the environment is not explicitly present.

Via the 3-tier Collective Mind scheme, the strategy steps of the Z-Model achieve a further extension for the team.

We illustrate this extension with a project example from the fictitious company Medical Fit. This company is a research-oriented pharmaceutical company. New drugs require a flexible research process, supported by innovative information technology. For this purpose, an innovative IT system is to be created.

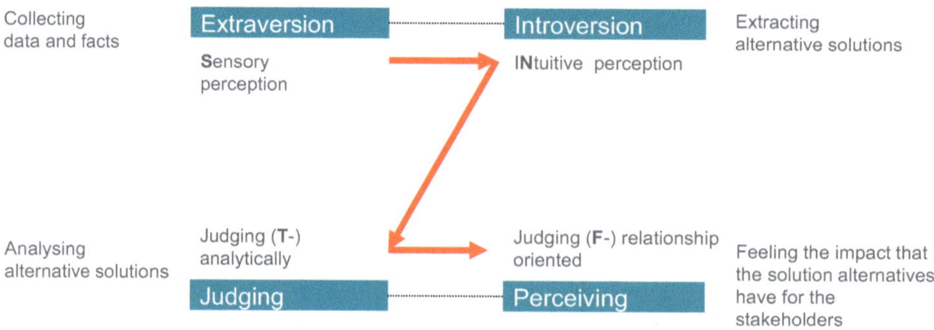

Fig. 4.13 Z-Model and strategy for dissolving biases

For better illustration, in Table 4.2, we anticipate the results of a textual representation of the target hierarchy.

A flexible research process means that new drug tests will be added and old tests will be phased out, which requires an adjustment of data structures and finally of the underlying IT systems. Here we focus on a central question: How flexible should the architecture of the process-supporting IT technology be? We will have an open discussion on this, since a recommendation is always dependent on the environment. We will especially focus on the acting on and eliminating of typical biases.

The project task now is: Find IT-architecture that supports a flexible and fast moving research process.

The example we discuss here is very likely a missionary project (see appendix "Fundamentals Diamond Model"), i.e. an innovative task, which affects many people. Ideally, solution finding should begin at the Goal-tier (after the information collection phase). In practice, there are two major threats: Creative ideas emerge and distort the course of the project in an undesirable direction. In addition, the Goal-tier is not systematically linked via the What-tier with the How-tier. And there is no conscious separation of these three tiers of abstraction. This can result in undesirable biases considerably hampering effectiveness, and at the same time, efficiency will suffer, because a lot of time and energy will be spent on details that provide no added value.

In Table 4.3 we have compiled typical biases for our example, which could potentially arise during the forming of the Collective Mind, as well as their possible characteristics, the risks they may cause and the necessary countermeasures.

To avoid the risk of biases, we propose connecting the strategy steps of the Z-Model with the 3-tier model of the Collective Mind (see Figs. 4.12, 4.13 and 4.14).

The leadership task of the project manager is to design this process systematically. In innovative projects, it is recommended that the number of team members with N-preference predominate (at first), so that System1 is particularly strongly represented in the team.

It is very likely that much creativity will result, but at the same time the risk of biases will increase. So it is always necessary to provide an environment in which System 1 remains in the background and System 2 takes the lead in correcting. This should not be carried out abruptly, by jumping directly to the How-tier, but the What-tier should firstly be built up. There, the Goal-tier will be split into aspects like processes (Table 4.3 in the case of the research process) or competencies (Table 4.3 in the case of experts) or components (Table 4.3 in the case of technology). These aspects should be designed in a

Table 4.2 Example of 3-tier model of the Collective Mind

Goal-tier	With the help of a system, researchers are given the opportunity to pursue innovative research paths flexibly.
What-tier	The system supports the main research processes and is designed as a toolbox with simple tools. Researchers can flexibly combine these simple tools to new tools.
How-tier	The tools serve the individual design of evaluation processes of all company substance data. For the creation of the tools, simple and widely known methods of research are used. However, the combination of tools is intuitive and is visually supported.

Table 4.3 Examples for biases in the Medical Fit project

Bias	Possible characteristics	Risks	Countermeasures
Priming	A specific technology or solution is again and again favored by dominant team members.	The solution does not match the requirements, but satisfies the innovative preferences of some team members. The original goal of searching for innovative solutions in research will be replaced by "use innovative IT-technologies".	Assessment of technologies according to objective criteria, search together with the dominant team members for their motives. Take decisions on technology later, use of decorrelating creativity techniques.
Over-prediction of events with a low probability	Postulation and insistence of the requirement that the IT architecture must be very flexible, because the research process could change dramatically at any time (Over-predicted event): Search for the "jack of all trades device".	Very likely the architecture is designed to be flexible, but this creates the paradoxical situation that small changes (frequent and probable!) of the research process cause substantial changes in the IT system.	Clarification of what "flexible" really means. It is to objectify how likely dramatic changes in the research process may occur. For this purpose, situations/scenarios should be developed.
Halo effect	An expert, who has been successful in other projects, is assigned to this new complex project, although they may lack the necessary business background.	The expert uses solutions that were successful in the past (according to their understanding of "innovative" and "flexible"), but no early review takes place and so the solutions used will not work.	Select experts on the basis of the required criteria and/or perform a transparent review by using the 3-tier Collective Mind model of the current project and the successfully completed project.
Anchoring	Focusing on a specific (perhaps more interesting) step in the research process (for example, early clinical testing) and neglecting the later stages, although the IT system supports the entire process.	In preparing the solution, the team unconsciously focuses repeatedly on the early stage of the research process.	Develop and apply verification criteria for the entire process, incorporate systemic views and simulate value-creating scenarios.

verifiable manner. Over the course of time, the environment of the What-tier will become more analytical-rational and will thus straighten out any biases. It is important that moderation of this tier is carried out by the project manager (or team member), who has had no predominance in System 1. The intention is that after a period of creativity a more rational phase will follow. The results of the review of the What-tier, could for example look like this: Analyzed, evaluated and documented main research processes, and IT-architectural sketch, containing an elaborate concept of future modules or formulated applications.

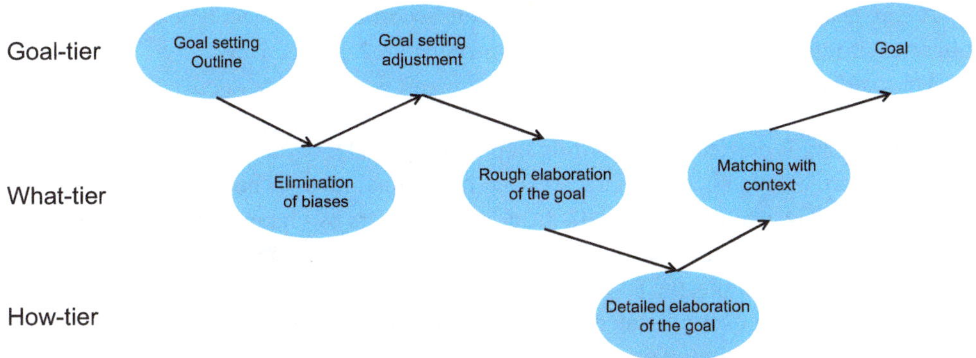

Fig. 4.14 V- and W-cycles for solution finding and elimination of biases in a missionary project

With respect to the over-prediction of events with low probabilities mentioned in Table 4.3, the analyzed and evaluated research process should include "probabilities" that are reliant on the organizational culture and allow an assessment as to whether a highly flexible IT-architecture is justified at all. If the research process changes in an evolutionary fashion, with the organizational culture subsequently allowing no abrupt changes, then a strong indicator exists for dropping the flexible architecture. For example, one solution could be the product fitting much better to the scenario: Flexibility on an evolutionary level is ensured, if the product allows the adaptation of modules, e.g. input screens. These modules are then updated when the research process changes. The result is a revised, bias-reduced Goal-tier based on a strategic decision.

Once the Goal-tier is sufficiently stabilized, detailed elaboration on the How-tier can start, based on the What-tier. The How-tier is dominated by System 2. For example, the focus here is on a detailed requirements document. These results need to be mirrored again on the What-tier. A rational and emotional check then needs to be made with all stakeholders, as to whether the results have the potential to be sustainably accepted. If they are acceptable, the solution should be re-mirrored with the Goal-tier to determine whether the target direction is still correct. If the solution is unacceptable, the root causes of this non-acceptance need to be clarified, including on the What- and How-tier.

The four-step strategy of the Z-Model, together with the 3-tier Collective Mind model, result in a model of V- and W-cycles for eliminating biases. The number of cycles and the combinations of the tiers depend on the convergence of the solution progress.

Each time the various tiers or cycles are passed through, the Collective Mind of the project team is better formed. In the end, the Goal-tier is sufficient to cause coherent associations with task and solution finding for all team members. The Goal-tier, therefore, is primed in the direction of the jointly elaborated project goal.

Let Us Now Turn to the Fourth Question Is it possible to weight the positive properties of System1 and System 2 according to project type and project phases, respectively, so that they can be applied appropriately?

A core task of the project manager is to create according to project type and the project phase environments that stimulate System 1 or System 2, respectively.

This can be done in several ways:

- Team setting is carried out according to project type [1].
- Since a project changes its character from its original idea, to implementation, to roll-out; it could be useful to cope with different characters by changing the team composition accordingly. In the Collective Mind Method [1, 19] we have outlined suggestions for this. If for example, we take an innovative project, in the concept phase, this type of project should be overpopulated (80%) by people with N-preference. At the beginning of the implementation phase, the proportion of people with N-preference needs to be successively reduced and the number of people with S-preference should dominate. This reflects the change in the character of the project.
- In the Walt Disney strategy we have already seen that different (project) phases should be supported by different roles (for example, the role of the project leader) and/or selective shielding of complexity in time and space. Even if it is not possible to change the structure of the team, it is essential that the project leader changes their leadership style during the project. In the concept phase, the leadership style supports creativity and intuition, so an effective search for the best goal and the best solution lies in the focus. During the course of the project the leadership style needs an increased focus on efficiency.

4.7 Stakeholder Management

... Life is always banal. Just now and then a pearl, a grain of sand, a shining particle, falls on our shoulder. And in the ocean of ordinary waves, power is the most banal vice among mankind. So why not draw a symbolic guillotine to indicate your power?

Fred Vargas, A Climate of Fear, English by Siân Reynolds

The term "stakeholder" covers all individuals, parties or groups interested in a project, i.e. having something at stake there. The object and purpose of effective stakeholder management is to proactively identify opportunities and risks that are associated with stakeholders and their interactions, as well as to shape and use them for the project. A project's character results from the project goal and purpose and both are related to the goals, interests and personalities of all stakeholders. Also keeping organizational circumstances in which stakeholders are involved, in mind, the result is finally a complex socio-technical system.

Stakeholders, along with their organizations, belong to a complex, and thus time-dependent (!) network of socio-technical success factors. At the same time, stakeholders apply different, probably time-dependent success criteria for the project ([1]). Within the framework of a project-specific PDCA cycle, the task is, especially for the project leader, to lead this system. We again emphasize here that "leadership" does not imply that the project leader will be provided with more knowledge than the others and therefore sets the direction. Rather, it is leadership for complex systems, which was described earlier in the book. Figure 4.15 shows a small, typical stakeholder network, consisting of members of the principal organization, as well as members of the agent and partner organizations. We have included "classic" stakeholder properties concerns/interests and influence/ power. In addition, we have also incorporated criteria objectives, and in particular, personality.

The network shown in the figure does not include any explicitly drawn connections (relationships); we will return to this later on. With this representation we want to illustrate that the various aspects of personality, objectives, influence and concerns of a person, as well as their role, mutually affect each other, like the atoms in a molecule. People are identified by their individual personality, but depending on the surrounding environment, which is caused by others, they may exhibit different behavior. Aspects such as concerns or influence can even change completely, depending on the environment.

Because different people may be involved in various other non-temporary or temporary organizations, these organizations also act via the stakeholders. By using the previous terms of micro- and macro-tier, various stakeholders are involved in various organizational macro-structures and the temporary organization "project", as an additional macro-structure, will be seen as an interference. At the micro-tier, a dynamic complex network of stakeholders forms that generates the character of the project on the macro-tier. This is then returned to the agents of the micro-tier and they produce the macro-tier self-referentially. This is the self-referentiality characteristic of all complex systems, whereby the "self" is the new temporary macro-structure "project". Stakeholder management can therefore consist of two completely different levels of intervention (please see the section "Transformation Management"): A micro-tier and a macro-tier. "Classic" stakeholder management usually only includes a micro-tier with the aspects concerns/interests and influence/power. We complement the modeling of the micro-tier by adding the categories success criteria, objectives, personality (MBTI, motives, values and basic assumptions), as well as attitude to the project (promoter, supporter, opponent and endurer). See also the discussion in the Collective Mind Method [1].

Figure 4.16 now shows an enlarged section from Fig. 4.15 with the personality flower specified for the project leader on the contractor side, a colleague from a partner company, and the principal's project leader.

In Fig. 4.16 the diversity, and subsequently the complexity, of interactions become obvious. In this section it is also made clear, how motives and values and their motivational schemes need to adjust to each other: Autonomy of the agent project leader lead, using the motivational scheme "I alone meet the requirements demanded by the client, because I always know best", will very likely violate three values of the principal project leader, namely, order, harmony and certainty. Also, looking at Partner X's colleagues, via the values "safety" and "trust", it is very likely that annoyances will be stimulated.

Conversely, the principal project leader may attach great importance to the value safety when creating the project product. The corresponding motivational scheme could be "I feel comfortable having documented the functions in black and white". If the agent project

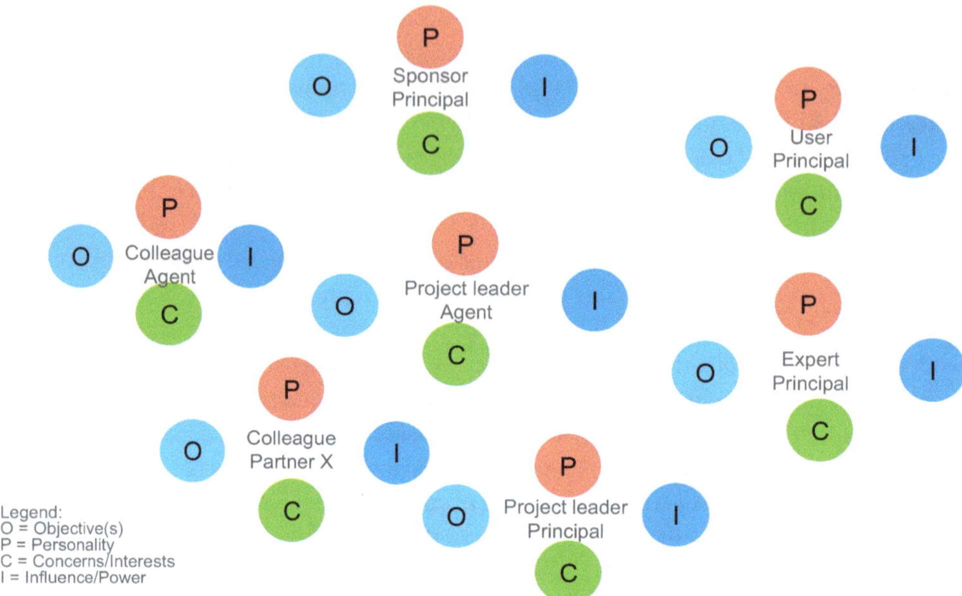

Fig. 4.15 Example for a stakeholder network

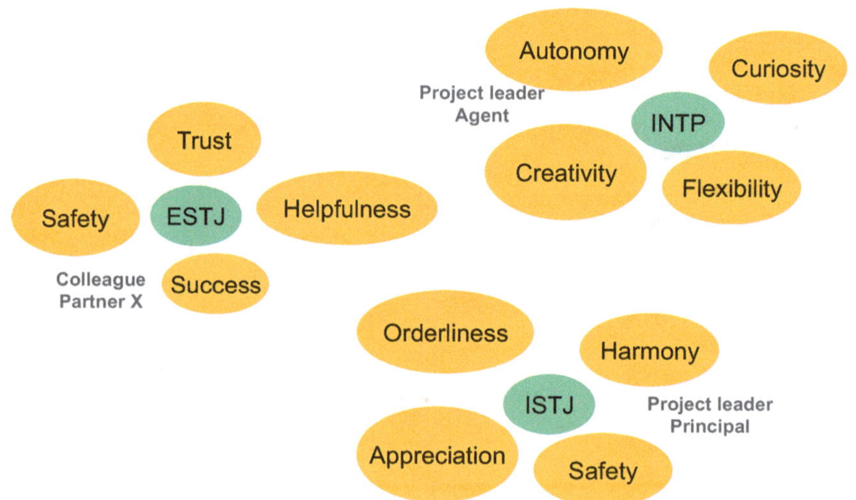

Fig. 4.16 Stakeholder management and personality

leader is unaware of their own preferences "curiosity" and "creativity" and does not respect the requirements resulting from the personality of the principal project leader, this will very quickly lead to corresponding conflicts.

Basic assumptions are also connected to motives and values. A principal project leader's basic assumption, which suits to their motivational scheme, could be for example: "A good product has proper project documentation". This could result in the product being rejected, despite no apparent malfunctions existing, and only documentation not meeting the requirements of the principal project leader. "Resonant adjustment" means that there is enough available space for personal motives, interests and goals in the team, so that each team member or stakeholder is willing to "subordinate" their motives, interests and goals to the Collective Mind.

As we demonstrated earlier (section "Fundamentals of Leadership"), with motives and values, depending on experience, different motivational schemes and assumptions are connected to each other. This results in different behaviors for the same motives and values, and creates a complex network of communication. Good stakeholder management makes these relationships transparent and takes appropriate measures. Measures that only address aspects of concerns/interests and influence/power, fall short, often much too short, because they cannot make real systemic relationships transparent.

Figures 4.17 and 4.18, each show an example of stakeholder management from consulting practice. For the sake of simplicity, motives, values and basic assumptions are not displayed here, but only the temperament preferences of some of the team members. As we will see, important measures for stakeholder management can be derived from temperament analysis. We recommend using such representation only for stakeholder analysis and not for the analysis of team conflicts in a team, as this may lead to considerable confusion and distortions, if, for example, the self-image and public image of involved team members do not match. In a "classic" social network, only the connections between the actors, their roles and sometimes the emotional relationships between stakeholders are displayed. Thus, social structures can be made visible, but in our experience, the reasons for "self-adjusting" social macro-structures are often hidden, because the personalities of the players are not considered.

Figure 4.17 shows an example of a network for a principal-agent relationship. The two project leaders are shown at the very top of the diagram. The agent project leader has an authoritarian leadership style (very extraverted, cares little about detail, with strong fact

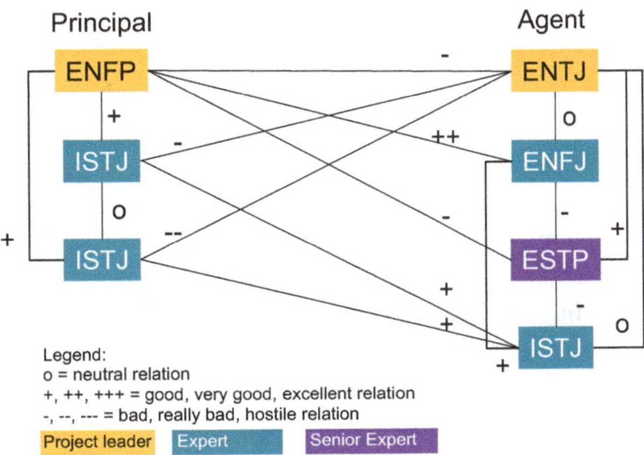

Fig. 4.17 Social network: principal-agent

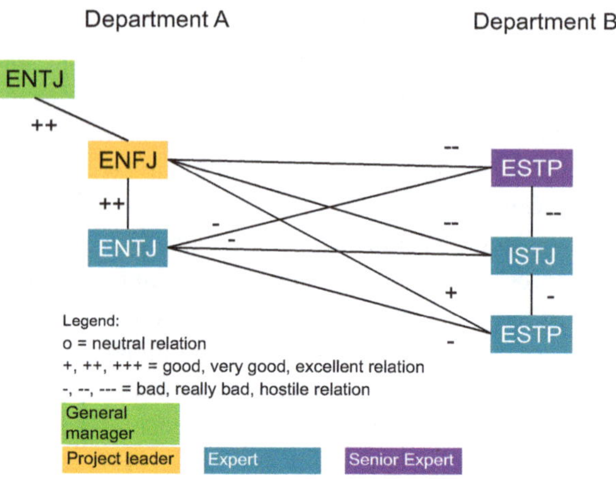

Fig. 4.18 Social network: department A – department B

and high goal orientation). On the one hand, this manifests itself in a bad relationship with all principal team members, and is also reflected in the poor relationship of the team members towards the company. The agent's senior expert (ESTP) represents another hurdle, because they have a poor relationship with all team members except the agent project leader. People with SP-preference use tools and methods aligned to their own preferences, which they handle with more and more sovereignty during the time. If individuals with SP-preference remain for a long time in the role of expert, this solidifies their unique selling points. The result can then, as in the current example, lead to the devaluation of other domain experts. In particular, this is so with the principal domain experts, with whom the senior expert has no relationship at all, due to their role as "superior" principal, and because "they are all below their professional level".

In addition, as can be easily seen from this example, this example supports the following hypothesis for the forming of micro- and macro-structures: "Similar people stick together" or in other words: To a large extent, the relationship network reflects interactions caused by personality preferences (especially with the inclusion of motives and values). Of course this can only be understood as a rule of thumb, and only serves as an initial orientation.

If the principal project leader has created the social network by themselves or with help of a coach, one possible measure could be the running of a joint team workshop on self-perception by MBTI and values analysis. This workshop could then be enriched by content processing of the project by revealing personal solution preferences and related basic assumptions using the Dilts Pyramid. Experience shows that after a few joint workshops, this significantly improves collaboration. If a joint workshop is not possible, the workshop could include just the principal (or the agent), respectively. In this case, the principal's employees (or those of the agent) will be trained to get to know their own preferences and align their behavior with resonant communication to the preferences of the agent's employees (or those of the principal). For example, for the agent project leader this could mean:

- Search for a common perspective and approach together with the agent expert ENFJ,
- take back their own extroversion (this of course depends on the degree of ENTJ extroversion),
- align the communication to the Big Picture, and less to the details,
- rational communication, yet simultaneously showing affection orientation, possibly supported by one of the ISTJs on the principal's side, whereas the level of detail for communication has to be agreed in advance between ENFP and ISTJ,
- scale down their own situation-oriented behavior (P dimension) and instead comply with a more stringent goal orientation.

These or similar measures are nota bene to be understand as hypotheses, which are to be checked in practice with respect to effectiveness. If one or more measures do not demonstrate effectiveness, i.e. whole team cooperation does not improve, this could imply that the assessment of personality preferences was incorrect, or that implementation was not carried out well or that other reasons are responsible for the relationships. In this case, an iterative PDCA process should start, accompanied by the coach, with the aim of finding the effective success factors and reviewing the related hypotheses.

Figure 4.18 shows an extract of the social network of two departments in a company who work together on a project.

In this example, a very good relationship exists between the managing director and the members of department A. Due to the personality preferences, this is not surprising. As we have learned above, this is based on a rule of thumb: Personalities with similar preferences have good relationships, because they recognize (unconsciously) behavior patterns that fit with their own patterns. This of course assumes that the perceptions are not interfered with by negative mutual experiences. Negative experiences that have been set in the mind, due to some common experienced situation, almost "always" encompass personality preferences.

The emerged macro-structure in Fig. 4.18, follows preference patterns even more clearly, in particular with regard to the relationships between members of department A, the managing director, and members of department B.

What is striking, is that in section B, the "climate" is completely negative. Just from the personality preferences shown here, this is not obvious. Further analysis based on values and basic assumptions, however, revealed that the senior expert shows a strong need for power, combined with a high need for autonomy. Their behavior displays an authoritarian style combined with very low willingness to engage in other professional ideas and solutions. Again, according to the first example, adapted measures should be initiated, and effectivity is probably significantly lower. It is likely that extensive interventions will promise more success (e.g. using the power of the managing director; working in tandem with colleagues who have a good relationship with ESTP; eliminating the employee from the project).

Both cases demonstrate quite impressively that this social network analysis provides insights that substantially exceed the potential of conventional stakeholder management, and therefore much more substantial and appropriate interventions can be applied to the complex system “project”.

En Route to an Agile Organization, Part 7

At the next meeting, Mr. Leibowitz summarizes the results of the previous workshops. The systematic application of the Collective Mind Scheme is used to agree on principles, which are the basis of the directive. “We will focus on the principles first. These are as follows:” Mr. Leibowitz goes over to the flip chart and writes them down. Right beside the principles, in parentheses, he then writes the expected values ”standardization” and “reproducibility” as order parameter and as control parameter “allowing some researcher freedom,” in order not to contradict the basic values of the department:

- The measuring instrument has to be calibrated according to DIN.[10] At least one measurement and repetition have to be carried out at a fixed time after a sample has been received. The point in time must be the same for all three test departments and must be coordinated with the research departments (Standardization: excludes departments’ individual actions. Standardized process sequence in all three departments: excludes time effects of the sample input and identifies outliers in the measurement).
- A measurement must be reproducible and understandable. The measured values and the measuring method are determined once a year by the governance unit together with all three test departments (Reproducibility: allows comparison of the measurements of the three departments).
- The measurement procedures can still be determined by the departments and their employees (maintaining freedom of research).

After Mr. Leibowitz has listed the principles, the room is cloaked in silence. He continues: “The principles of the guideline – I am no longer talking about a directive, but a guideline – shape the above system parameters and are, next to the governing unit and their orders, important system parameters. Nothing more! Therefore, this is a guideline and not a directive, as it goes far beyond the simple instructions of the original directive. This should be implemented in the departments, based on existing best practices. There are likely to be only minor changes in the procedures within the test departments. The guideline, therefore, merely closes the gaps, without fundamentally altering how the colleagues concerned function. Extremely efficient! In addition, the last system parameter is that from now on the governance department will carry out audits according to the guideline, whose results must be submitted to the management. This is necessary to guarantee the efficacy of the guideline.”

Ehrlich continues: “I believe we can present this result to Mr. Freschi, because our mental model matches his mental model: As a successful research director, he should not act against the basic values of his unit, whilst at the same time, he also has to meet

[10] DIN is the German Institute for Standardization.

scientific requirements, namely standardizability and reproducibility of measurement procedures. I am confident that this is so, and it is highly likely that we can include Mr. Freschi in our Collective Mind. And most important of all: We have now taken into account all stakeholders in their environment, so in this respect no complexity should be expected. The department cultures have subsequently been taken into account almost automatically; nevertheless, the interactions between individuals and departments should be examined more closely, in order to analyze and stabilize adaptation of the working methods according to the new guideline. And finally, the roll-out of the guideline means a change, a transformation, if only a small one, and should be carried out professionally," he concludes with a wink.

4.8 Interaction Between Micro- and Macro-tier

Elisabeth:
...
Under what title shall I woo for thee,
That God, the law, my honour, and her love
Can make seem pleasing to her tender years?
King Richard:
Infer fair England's peace by this alliance
Elisabeth:
Which she shall purchase with still-lasting war.

William Shakespeare, King Richard III

In the examples for stakeholder management we have seen that even the use of only one model (i.e. the MBTI model) for modeling the success factor "stakeholders" is of significant value in understanding interactions in a social network.

In this section, in addition to the success factor "stakeholder", we take the success factor "organization" and extend our models to organizational interaction with regard to projects. Furthermore, we also consider the interaction of individuals and organizations. We focus here on the Dilts Pyramid for modeling the levels of identity (with temperament model MBTI [20, 21]), as well as values and basic assumptions (with Schein's culture model [22]). We assign personalities to the micro-tier and organizations with their interactions to the macro-tier. We follow our earlier publications [23, 24] and expand them here accordingly.

Considering the micro- and macro-tier, as well as their interaction, is important for many applications and opens both a systemic and systematic approach for directed intervention. The integrated model presented here can be applied in very different areas:

- An interdisciplinary project, carried out by several organizations.
- Collaboration between client and consultant.

- Collaboration of two (or more) departments (for example different business departments).
- Collaboration between department and assigned sub-departments.
- Transition of an employee from one organization to another (for a specific time period after the transition, an interaction occurs between two cultures: The interaction of the new and the previous cultures).
- The merger of two companies or the takeover of a company by another one.

If the collaboration in these examples is not "organized" by a Collective Mind, the persons involved will primarily act and feel as if they are members of their parent organizations. – Even if a Collective Mind exists, there will always be situations in which behavior manifests itself, which will only become clear through a deeper understanding of the parent organizations. A lack of knowledge of these relationships makes intervention pure chance, and far from professional. In the worst case scenario, blockades will be generated in collaboration, with no one in the team aware of how these blockades are created by the team itself.

For the sake of simplicity, we will restrict ourselves below, to the interaction of two organizations.

Assuming that the behavior of organizations is essentially determined by people, i.e. by their temperament, the idea is to transfer the term "temperament" to an organization. For this reason, Bridges [21] has applied the MBTI model to organizations and defined an organizational temperament.

Subsequently, each organization has a temperament that is generated by the temperaments of its members. It may be that an organization's temperament is originated by a few people (the founder of the organization or (former) executives) or has evolved over time in accordance with the vast number of temperaments of people within the organization.

The interaction between organizational temperament and the temperament of an individual was first approached by Bridges himself. If we start describing interaction in a simplified (!) way using the 16 typical organization temperaments and the 16 typical individual temperaments, the result will be a 16 × 16 interaction matrix.

Even with this simplification, and further assuming stereotyped behaviors, the result has a range, which can no longer be handled by practical processing in the PDCA cycle. To make biases visible, we apply the rule of thumb: "A simplified, explicit model is better than no explicit model". We further simplify and only consider the main personalities of Keirsey's temperament model [25], namely the Guardians (SJ), the Artisans (SP), the Rationals (NT) and the Idealists (NF). This results in a manageable 4 × 4 matrix (for further information please see the main personalities (function pairs) in the appendix "Fundamentals MBTI").

These main personalities are described by the manner in which they use language and their use of models, methods and tools. The use of language is characterized by the attributes "concrete" or "abstract" and the use of models, methods and tools is described by the attributes "cooperative" or "use-oriented". An individual or an organization with NT temperament will prefer abstract language and adheres to the use of those models, methods and tools they find suitable.

The 4 × 4 matrix (horizontal: individual, vertical: organization) shows the interaction between individual and organization and their 16 possible combinations. It is striking that the interaction between individual and organization is not symmetrical:

- On the diagonal, individual and organizational temperaments match, so we indicate this with an image of the sun.
- Due to their characteristics (concrete language and cooperative behavior), an individual with SJ temperament is able to get along with organizations that have SP, NT and NF temperaments. We characterize this with a red check sign, indicating that this interaction is running reasonably well, but not optimally.
- SP temperaments have a slightly bigger problem in adapting, when their use-oriented behavior does not match the use-oriented behavior of NT organizations, and this is indicated by rain clouds.
- Due to their characteristics, i.e. abstract language and use-oriented behavior, in almost all organizations, NT temperaments struggle, especially in SJ organizations, and this is indicated by a thunderstorm image.
- Employees with NF temperament will struggle in an organization with use-oriented behavior (NT and SP temperament), since this is in opposition to their cooperative behavior.

In the above, we have neglected effects of learning and development for individuals and organizations: Learning here means mutual interference in the behavior of an individual and an organization over time. We illustrate this by explaining the interaction of an NT temperament in an SJ or SP organization (abbreviated to S organization):

The strengths of S organizations lie in the efficient implementation of tasks and concepts. The strengths of NT temperaments lie in the development of ideas and concepts. The conflicts illustrated in Table 4.4 between NT temperaments and S organizations can then be eliminated if NT individuals, who think in new concepts, communicate in a

Table 4.4 Interactions between individual and organization

Individual/ organization	SJ (concrete, cooperative)	SP (concrete, use-oriented)	NT (abstract, use-oriented)	NF (abstract, cooperative)
SJ (concrete, cooperative)	☀	✓	⛈	✓
SP (concrete, use-oriented)	✓	☀	🌧	🌧
NT (abstract, use-oriented)	✓	🌧	☀	🌧
NF (abstract, cooperative)	✓	✓	🌧	☀

fact-oriented way to the S organization. In the language of the Collective Mind, this means that the NT individual must transfer their concepts from the Goal-tier, to the What-tier or even better to the How-tier.

An NT individual therefore, has to elaborate their ideas, concepts and solutions in more detail than is usual for NT temperaments. This is an active learning process for an NT individual. The organization can support this process through specific language and a desire for details, by asking the NT individual specific questions. The S organization then takes up these ideas, concepts and solutions and implements them directly. This is an active learning process for the S organization. The S organization can learn that an NT individual is wired differently to the S organization, struggling detailing and implementing ideas and concepts. However, this may be outweighed by their contribution to the development of the S organization. For this purpose, the S organization must be able to recognize the use of the NT individual by self-referencing. However, this is not always the case: In the case of the S organization having "difficulty in learning", the NT temperament will eventually leave the organization.

Using the temperament model for individuals and organizations, we have revealed their interaction by way of an example. We are aware that the above interaction model is significantly simplified. Please feel free to apply the model, and you will then perceive what impact the new insights will have for the interaction between you and your organization.

However, this in itself is not sufficient to adequately describe the interaction between individuals and organizations: Both organizations and individuals have values, motives and basic assumptions, which interact mutually. On the one hand, individuals bring in their motives, values and basic assumptions, and on the other hand, collective experiences emerge within an organization that lead to new collective values and basic assumptions. These values and basic assumptions exist for the most part only (unconsciously) in the minds of members of the organization and will be subsequently transferred to new members. This is reflected in the well-known effect that a new employee (although technically sufficiently competent) is unable to understand an organization or even rejects it; however, if the individual remains in the organization, in time, they will adopt and live its culture.

We can go one step further, and for the description of the interaction between individual and organization, we can add their respective values and basic assumptions. Values and basic assumptions of an organization are described by cultural models. In the literature [26] there are different cultural models that refer to all cases known to us, with respect to values and basic assumptions and related behavior patterns. In our experience, v-memes from Spiral Dynamics [7] are excellent to use as a master model of values and basic assumptions (see appendix "Fundamentals Spiral Dynamics"). For this reason, in the following, we consider Spiral Dynamics as a reference model for cultural models. However, specific cultural models still have their role, as they provide specific, operationalized perspectives on v-memes.

Hofstede and Hofstede [27] were the first to operationalize the concept of culture for different countries in a cultural model. In an analogy to the personality models, they introduced five dimensions ("power distance, individualism, masculinity, uncertainty avoidance, long-term orientation"). Furthermore, they measured the majority of countries in the world with respect to these dimensions on a scale of 0–100. The dimensions defined by them cannot easily be assigned to v-memes, but contain per dimension, a mix of different v-memes and associated behavior patterns.

Schein [22] applied the idea of culture dimensions to organizations and identified six main dimensions. Schein uses the following definition for an organization's culture: "*The culture of a group can now be defined as a pattern of shared basic assumptions that was learned by a group as it solved its problems of external adaptation and internal integration, that has worked well enough to be considered valid and, therefore, to be taught to new members as the correct way of perceive, think, and feel in relations to those problems.*"

The six main cultural dimensions, according to Schein are: Reality and truth, time, space, human nature, human activity and human relationship. Schein's dimensions contain different facets (sub-dimensions), some of which are not always characterized consistently by opposite pairs. It lacks a continuous measurement scale and measurement specification for dimensions, and therefore, operationalization as referred to by Hofstede and Hofstede is not directly possible.

In [23] and [24] we formulated Schein's dimensions consistently over a selection of pairs of opposites, aware that this was a simplification. But, as is often the case, we also learned that this simplification will help to produce a sufficiently high variety for practice, which makes it possible to intervene successfully and comprehend complex interactions between organizations and individuals.

We described the facets of Schein's dimensions using our selected pairs of opposites and related characteristic questions. In addition, we assigned facets to the v-memes of Spiral Dynamics. This assignation acts as an introduction to a higher level of abstraction, which operates as a rule of thumb and thus helps to form intuition for cultures.

For example, we assigned the dimension "reality and truth" with the pole "absolute truth" and the blue v-meme (basic need "orientation and control", v-meme: "The level of existence, which is subordinate to absolute truth and where lasting peace is sought: absolutist, conformist, religious"). This assignation is to be understood in such a way that we connect the major part of this facet with the v-meme, but other v-memes can also be included in this facet. So for example, there may be justification for assigning to the pole: "absolute truth" fractions of the red v-meme (basic need "self-appreciation and self-protection", v-meme: "The self-determined, "heroic-life-without-consideration-of-others" level of existence: egocentric, hedonistic").

Space

Personalized space (green v-meme) or depersonalized space?

Will a personal space order and design be supported or is the space for the common good?

Demand-oriented space (yellow v-meme) or hierarchy-oriented space?

Will the interior design support the required demands or the manifestation of hierarchy?

Reality and truth

Objective (blue v-meme) or relative?
Will the truth be regarded as relative and therefore depend on the observer or situation or environment?

Relationship-oriented (green v-meme) or task-oriented?
How will reality and truth be found? Will it be checked by pragmatism via trial and error, or will a scientific approach (model and measurement be used)? Will power, expertise be decided democratically? What levels of truth are there? To what extent is personal truth and reality accepted? How are decisions made: relationship-oriented or task-oriented?

Time

Polychrome (orange v-meme) or monochrome time?
Will tasks or activities be worked out sequentially (monochrome time understanding) or synchronized in a time unit? Is attention in a time unit focused on a person, task or situation or on several of these?

Planned (blue v-meme) or developing?
Will time be planned based on clearly defined goals and milestones on the basis of an objective reality or will the developing process determine the results in a natural manner?

Human Nature

Intrinsic (green v-meme) or extrinsically motivated?
Are people motivated by themselves and therefore able to take responsibility and make decisions, or do they need external stimulation?

Mutable human nature (green v-meme) or fixed human nature?
To what extent does disposition determine skill? To what extent is a human able to develop?

Human Activity

Activity-oriented (orange v-meme) or condition-oriented?
There is a prevailing view that each individual is responsible for their own fate and therefore only has to do the right things right, to take fate in their hands (American slogan "We can do it!")? Or does a certain amount of fatalism exist, coupled with the insight of inevitabilities caused by set conditions of the environment?

Systemic (yellow v-meme) or purely functional
Will actions be aligned by reflection based on model orientation, on interconnectedness and the dynamics of the respective situational environment, or will actions be determined by purely functional needs?

Human Relationships

Collectivism (blue v-meme) or individualism?

Will the alignment of standards and rules follow the community or the individual? What is more protected: individuality or the interests of the community? Is community merely a means to achieve a goal?

Large power distance (red v-meme) or small power distance?
To what extent will hierarchy, formalities, rules and standards be emphasized?

En Route to an Agile Organization, Part 8

After research director Herbert Freschi, quickly and unbureaucratically agreed with the proposal and subsequent course of action, Priesberg, Ehrlich and Dan Leibowitz meet up in order to discuss further the issue of stakeholder management. The formulation of the guideline by the organization and embedding within it makes it likely that only a fine adjustment in terms of stakeholder management will be required.

Ehrlich begins: "It is now extremely important to strengthen the positive momentum for Mr. Wallace, Ms. Snyder and Mr. Broad, and use it for department-specific implementation of the directive."

"I think this is best done by involving Mr. Wallace in the last point of the guideline, which concerns the measurement rule. His own procedure should play a role there or at least provide the starting point. We already have Ms. Snyder on board, as her department already provides the best results and calibration of apparatus according to DIN[11] will be no great hurdle for her. Finally, Mr. Broad finds the directive almost "anti-Tayloristic," as he euphorically assured me in the aftermath of the last meeting. He has distanced himself from his department until now; the guideline offers him a great opportunity and will therefore be easy for him to get those colleagues affected behind him. However, he should be careful not to overdo things, so that he does not lose his standing in the department. For this purpose, a detailed analysis of the interaction between him and his department will be necessary to elucidate the risks for him and control them. This should be the next step," Leibowitz adds.

Priesberg brings the review to an end: "The final step will be change management, meaning the roll-out of the guideline by means of adapted best practices. In this respect, I believe I have understood the following: Auditing has to be carried out cautiously. If it is not carried out, the guideline will be ineffective. If it is carried out too aggressively with subsequent sanctions, acceptance will also disappear."

Ehrlich is lost in thought and smiles, "I think I have an idea …."

To illustrate the interaction of organizations and individuals, we will consider the interaction of a business department and an IT department of a fictional pharmaceutical company. We have stereotyped the two departments with opposing views, without restricting (in our opinion) the generalization here.

[11] DIN is the German Institute for Standardization.

Figure 4.19 outlines the example scenario (in the example we omit the dimension "space", since it is not relevant here):

We have characterized our example as follows:

The IT department and the business department are working on a joint task.

The fictitious pharmaceutical company is global. The business department and the IT department are located in the same country. We therefore do not need to consider country specific culture here.

The temperament of the pharmaceutical company will be characterized by type ISTJ: It has a strong inward view of its own products. While acting as a research-based pharmaceutical company, skills dominate, having been acquired over several decades: Active ingredients are identified via defined procedures, they are tested and developed and finally officially approved and marketed.

The business department is a drug regulatory affairs department, which operates drug approval according to legal requirements. As an entire company, it has the temperament ISTJ.

The IT department sees itself as a global partner of the business and with their IT systems, supports all other departments in the company. It is a flexible and innovative department. A few years ago, the department was absorbed, during the acquisition of an IT company, and settled independently within the pharmaceutical company. It has preserved its original ENFP temperament and because of this, is seen as an island within the company.

The business and IT departments will create a small IT system, which complements the existing IT landscape of the business department. The business requirements for this small

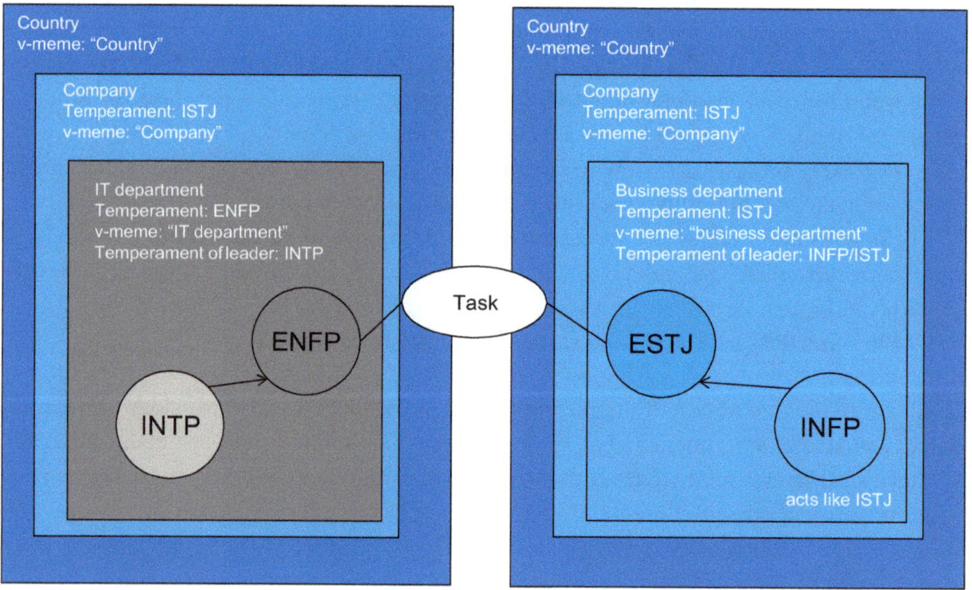

Fig. 4.19 Interaction structure between IT department (left) and business department (right)

system are easily recognized, as they are restricted by governmental conditions. Because of these conditions, the task is not carried out as a project, but instead as ongoing work ("on call") between these two organizations. We have assigned the project type "carpenter project" to this task (the project therefore has the temperament IS__, see appendix "Fundamentals Diamond Model"): The number of stakeholders is small and only recognized methods and models for the business or the IT department will be used. This corresponds to a low degree of mission (novelty), as well as a low level of abstraction (complexity), and finally a low level of innovation (technology).

We have typified the IT department as follows:

Mission The IT department sees itself as a global partner of the business, who wants to shape it efficiently and effectively by IT systems.

Strategy The IT department provides an innovative project portfolio driven by the requirements of the business.

Semantics The understanding of mission and strategy are expressed in the choice of the wording: They speak of "solutions, applications, projects and services".

Organizational Temperament The temperament of the IT department has the signature ENFP:

- **Extraverted:** The IT department has open borders, gets advice from the outside, and sticks to the motto – "The solution is out there, we just have to find it".
- **INtuitive:** Runs to high performance when it comes to the big picture; is a little careless with routines, emphasizes paradigm shifts.
- **F**eeling: Makes decisions based on value propositions, meaning that one of these becomes more effective by support.
- **P**erceiving: Looks for opportunities and wants more information, sets general standards, works open mindedly and tolerantly.

V-Meme Culture The culture is characterized by the following characteristics of Schein's culture facets. (Here again, the rule of thumb we discussed in the assignment of personal temperaments and values is valid: "Values and temperaments fit predominantly well with one another"):

- Relative truth
- Relationship-oriented objectivity
- Polychrome time
- Developing time
- Intrinsically motivated
- Mutable human nature

- Activity-oriented
- Systemic
- Individualism
- Small power distance

We have typified the business department as follows:

Mission As a service partner, all information required for the approval of an active ingredient is made available at the requested date to the authorities. The business department is thus a dedicated service provider with a dedicated task: Preparation and execution of product licensing, taking into account legal conditions.

Strategy Each employee provides their contribution in a well-defined and experienced process for approval. The spectrum ranges from toxicological studies, and statistical evaluation of data, to the preparation of the dossiers.

Semantics The mission understanding and strategy are expressed using the following semantics: The service is offered as a product, the dossier.

Organizational Temperament The temperament of the business department has the signature ISTJ:

- **I**ntroverted: Has a closed border, adheres to the motto – "We have the solution; we just need to find out how we do it"
- **S**ensitive: Is very strongly engaged when dealing with details, prefers a solid routine and a step-by-step process of change.
- **T**hinking: Thinks in terms of rules and exceptions, appreciates logical thinking, is a "social machine".
- **J**udging: Sets clear and specific standards, works towards results.

V-Meme Culture Culture is characterized by the following characteristics of Schein's culture facets:

- Objective truth
- Task-oriented objectivity
- Monochrome time
- Planned time
- Extrinsically motivated
- Fixed human nature
- Condition-oriented
- Purely functional
- Collectivism
- Large power distance

The following Fig. 4.20 illustrates interaction with respect to the setting of organizations:

IT employee Fred Landau, with ENFP temperament (champion), and business employee Sally Harris, with ESTJ temperament (Supervisor), work together on the carpenter task, with ISTJ temperament. Sally Harris is responsible for the execution of the task.

As a result of his long-standing organizational affinity, Sally Harris' line manager, who has an INFP temperament (Healer) displays behavior, which is strongly characterized by the organizational temperament of his organization, i.e. ISTJ temperament.

Fred Landau's line manager, who has an INTP temperament INTP (Architect), is able, to a large extent, to develop, within his IT organization. Although from time to time, his analytical decision-making style causes resistance in the relationship-oriented organization.

Sally Harris and Fred Landau, two colleagues in the pharmaceutical company, are supposed to tackle a carpenter task. Neither Fred Landau, as an IT employee, nor Sally Harris, have the appropriate temperaments for this task. However, whilst the temperament of the IT department does not correspond to the task, the temperament of the business department does. Also, in the IT department, such a small task simply falls by the wayside, because this department with the innovator mindset puts the focus on "bigger" things.

Therefore, Ms. Harris and Mr. Landau are together unable to find the right tempo to tackle the task: Fred Landau does not take the task seriously, because he considers it too dull and in his department such tasks are often not taken seriously either. He turns up late for meetings, changes the time of meetings, and is always proposing new technologies to

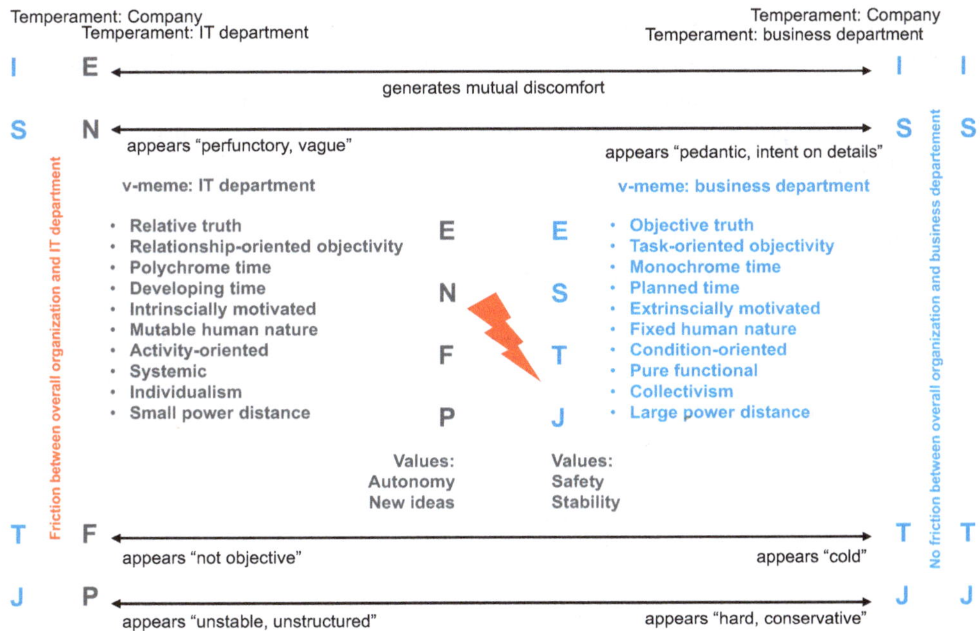

Fig. 4.20 Interaction between IT department (left) and business department (right)

solve the problem. Individual temperament and departmental temperament mutually reinforce each other.

Fred feels slightly encouraged by his department not to take this small task seriously. Most likely, this task is not even in his target agreement. Sally tries to steer him using ESTJ temperament and remind him of his duties. She is also strengthened by her department's temperament.

This often results in ineffective escalations involving line managers: Sally's manager only expects a simple implementation of the business requirements and interprets the problems in the team as technical incompetence of the IT side. Fred's manager supports him in his analysis of the project situation, but is ultimately unable to help, because he lacks the necessary methodological toolset.

The battle fronts are emerging and the project which has already been launched with elan by the business department, is beginning to emerge in structures characterized by individual and organizational temperaments and characteristics formed by cultural facets.

In the following, we outline potential manifestations of cultural facets with regard to collaboration based on Schein's cultural dimensions. At the same time, we show which cultural facets are supported by which individual and organizational temperament manifestations (the cultural dimension "space" is not relevant for our example and is therefore not included):

IT department: **Relative truth**
Business department: **Objective truth** (v-meme: blue)

For IT, truth is strongly linked to individual perspective and the given situational context. The business department however, assumes that there can only be one view of the truth, namely its own. This cultural facet is amplified by respective individual and organizational temperaments. Since the task has an ISTJ temperament, the business department which has the same temperament, feels that its attitude is recognized and sees itself as a "goal-oriented maker".

IT department: **relationship-oriented objectivity** (v-meme: green)
Business department: **task-oriented objectivity**

The business department cultivates task-oriented objectivity, while the IT department weighs its decisions based on relationships with stakeholders. This facet is supported by the T-temperament or F-temperament of respective organizational temperaments. This interaction amplifies the blocking effect of the relative truth/objective truth facet.

IT department: **Polychrome time** (v-meme: orange)
Business department: **Monochrome time**

The business department expects the IT department to focus its time on the current project. The IT department, on the other hand, sees it as "normal" their employees' available time is divided between several projects and tasks. Discussions on resource planning are inevitably unfruitful. The value-destroying complexity of such interaction is sustained by the P-temperament of the IT department and the J-temperament of the business department.

IT department: **Developing Time**
Business department: **Planned time** (v-meme: blue)

The business department is convinced that a fixed goal with fixed milestones is the basis for efficient and effective collaboration. The innovative IT department has experienced that innovative solutions require time to mature and transfers this knowledge to all types of tasks, even operational ones. This cultural facet also reflects the contrast between _N_P and _S_J temperaments: An open, sometimes unstructured and innovative _N_P attitude compares with a clearly structured and firm _S_J attitude. In the IT department, milestones are rarely reached on time, because when alternatives arise, they will be (unconsciously) integrated into current activities.

IT department: **Intrinsically motivated** (v-meme: green)
Business department: **Extrinsically motivated**

The IT department sees itself as a designer of the business and acts proactively. The business department feels like a responsive service department and feels patronized by the IT department. This form of interaction is sustained by the EN __ temperament of the IT department and the IS __ temperament of the business department.

IT department: **Mutable human nature** (v-meme: green)
Business department: **Fixed human nature**

The IT department adheres to the view that an employee's ability to develop on the basis of their talents should be embraced. Depending on current organizational skills, the manager helps employees analyze the project situation and implement appropriate patterns of action (this of course requires knowledge of the influence factors mentioned here). The business department has no development-oriented leadership style and only conveys the behavior desired by the organization. This interaction is supported by the _NF_ temperament of the IT department and the _ST_ temperament of the business department.

IT department: **Activity-oriented** (v-meme: orange)
Business department: **Condition-oriented**

This dimension is very much related to the "intrinsic-extrinsic motivation" dimension: The IT department relies on its creative power, while the business department feels embedded within the company and operates within the framework provided by the company.

Transferred to individual employees, employees of the IT department act independently, i.e. on the basis of a mandate which they have received for a project or task from their manager and their organization. With respect to decisions, business employees need constant confirmation by their organization. This interaction is sustained by the EN __ temperament of the IT department and the IS __ temperament of the business department.

IT department: **Systemic** (v-meme: yellow)
Business department: **Purely-functional**

The business department sees itself as a provider of functions or products, there is no need for a cross-department or system-wide view. Necessary activities will just be carried out. In carpenter projects or tasks, the business department is clearly at an advantage by following this procedure, but it struggles when dealing with innovative tasks. On the other hand, the IT department overshoots with simple tasks and requires a more comprehensive view: It tends to expand the horizon with ISTJ-tasks, making them bigger than they actually are. This interaction is sustained by the NF temperament of the IT department and the SJ temperament of the business department.

IT department**: Individualism**
Business Department: **Collectivism** (v-meme: blue)

The business department understands itself as a product supplier; the individuality of single employees has no role to play role here. Decision-making competence lies with the organization with appropriate bodies. The IT department's individuality-oriented culture is therefore perceived by the business department as non-consensual. The constant querying by business department employees to their managers is incomprehensible to employees of the IT department who believe this hinders cooperation and is detrimental to the progress of a project. In our experience, such interaction cannot be assigned to a specific temperament dimension, but emerges through the actions of the whole temperament in conjunction with experience.

IT department: **Small power distance**
Business department: **Great power distance** (v-meme red)

A greater level of power dominance in the business department leads to visible signs of this in the spatial design and to a strong indirect influence in the project. The line manager of the business employees is almost always present at project team meetings: He is able to reject approved project decisions at any time. This interaction is strongly induced by the task temperament, which favors or even necessitates a hierarchical structure.

On the basis of the above example, the following statements can be made:

The temperaments of individuals fitting the temperaments of a task are – in addition to technical expertise – the central variables influencing the success of a task.

Temperament and culture of the involved organizations are social fields that make team building more difficult. If the project or the task does not develop its own organizational temperament and its own organizational culture, then there will be no adequate differentiation from the environment: The temperaments and the cultural manifestations of the organizations involved will have a great influence on the interaction of individuals. Individuals thus transfer organizational characteristics of their home organization into the working group. If their temperaments and the cultures are different, as shown in the example above, this results in blockades to collaboration.

What measures should be taken to ensure that such collaboration will be finally successful? We propose the following complementary, potential interventions, whereas a forward-looking cost-benefit analysis will likely limit the number of possible inventions:

- The simplest and most effective way is to replace Fred Landau by a similar technically competent colleague who is closer to the ISTJ temperament of the task. Sally Harris already has the suitable temperament (ESTJ), so the task is likely to be finished successfully, if there are no hostilities on a personal level.
- The working group is accompanied by a project coach, who is able to identify the influence of temperament and culture, and makes Sally Harris and Fred Landau aware of this, in order to stimulate team building (see also section "Virtual Team" in chapter "Consequences for Management Systems").
- Fred Landau and Sally Harris' line manager is aware of the influence of temperament and culture and leads their employees accordingly. In our experience, in most cases, this is a time-consuming and lengthy solution, which appears to be less efficient and effective for dealing with a short-term task.
- The task is treated as a real project, so that there is a team to shield the "negative" influence of the origin organizations (complexity is shielded in space and time). However, this is not sufficient if individual temperaments do not match the task. Also here, we recommend coaching by a project coach, who has a suitable temperament and is a member of the team.

4.9 Transformation Management

The weight of this sad time we must obey;
Speak what we feel, not what we ought to say....

William Shakespeare, King Lear

En Route to an Agile Organization, Part 9

The three key stakeholders, Mr. Wallace, Ms. Snyder and Mr. Broad, have invited all fifty colleagues who will be affected, from the three test departments and the relevant research departments, to a kick-off meeting: This is the start of upcoming change management, which is tied to the introduction of the guideline. Right at the outset, the choice of words is a deciding factor for the success of any change. If the "wrong tones are struck", the result might be completely different, despite good intentions, which is typical of self-referential systems.

Priesberg, Ehrlich and Leibowitz are not involved, because they are management. At this event, which is similar to a collegial case consultation, the aim is to win over all those colleagues who will be affected; a "bottom-up" strategy for collaboration. Of course, the new governance unit and its role is already providing "top-down" support, but that should not be the focus, especially at the beginning.

Broad opens the event. He learned from the previous interaction analysis that he should not distance himself too much from the values of his department. In order to prevent this, he moderates the event and factors in the role of the management in a well-controlled manner:

"Dear Colleagues, unfortunately, a while ago, we noticed that our three test departments are producing different results for identical samples. Ms. Snyder, Mr. Wallace and I have looked at the whole thing and found out that our three departments have drifted apart. So, a little correction is needed to move closer together again. He is interrupted by a question: "And are you able to just do that? Has management no interest in this? That would be something new!" Murmuring can be heard throughout the hall.

Broad continues: "On the contrary, management is very interested…" He is interrupted again: "Oh, then everything is clear." The heckler waves his hand dismissively.

Broad, now becoming increasingly upset, proceeds: "Could we please discuss the facts first? Yes, management is interested. It would be bad if that were not the case, because that's what they are paid for. But they have allowed us to determine how to standardize the measurement procedures ourselves. Imagine, if you had to apply endless regulations in great detail – would you enjoy this?" Now, a laugh ripples through the crowd. "You see, therefore we tried a different approach."

Ms. Snyder walks over to a large blackboard where the principles of the guideline have been clearly chalked up and awaits participants' reaction. At first, there is nothing reaction, but then a whisper can be heard. "Is that all?", asks a young colleague. "Yes, that's all," replies Ms. Snyder. "So that means we can continue to apply our specifications?", the young colleague continues. "Yes, with two minor differences: The specifications must of course be somewhat revised. For example, with respect to calibration, time of measurement and quality of the documentation", Ms. Snyder answers patiently. "But the biggest deal is surely yet to come," interrupts a more mature colleague. Mr. Broad returns to take the reins: "I'm not sure if we should refer to it as the 'big deal', but it is a fundamental change: After the introduction of the guideline, measurements will be audited at random by the unit which has issued the guideline. And before you all stand up in arms… this offers us a very decisive advantage, in contrast to a 'senseless' rule. Does anyone have any idea what that is…..?"

After a long silence, which is interrupted by occasional murmurs, the young colleague once again replies: "We now have the opportunity to improve our specifications by, presumably via you …." He points to the three key stakeholders: "By giving feedback. Since an audit certainly gives rise to feedback."

"Thank you, that is exactly what we are hoping for. We will then improve as test experts and our results will also be much more visible because we are in reality like mice working away in dark cellars, and a little bit of daylight will do us some good." concluded Broad, now hoping that the skeptical mood will turn into a more positive one. And it does. The rest is just routine: Best practices are adapted according to the guideline and all colleagues are subsequently inducted. The three key stakeholders, Mr. Wallace, Ms. Snyder and Mr. Broad, will again be in charge of the project and will remain in close contact with Heiner Priesberg, the head of the new governance department. Mr. Freschi is now firmly convinced that it is the right decision to propose and support Heiner Priesberg for this role, and with his colleague Ehrlich in the background, he will certainly be able to take on further challenges. And with great curiosity, Freschi harnesses the creative field that Ehrlich and Priesberg generate, but prefers to keep that to himself……

As we have seen above, social complexity is nothing more than communication. A dynamic macro-structure with characteristic patterns emerges, which shows its effects as a social field or social system.[12]

This social field can be value-creating ("the whole is more than the sum of its parts") or value-destroying ("the whole is less than the sum of its parts"). Usually, the transition from value-creating to value-destroying is not a gradual one, but slowly infiltrates an organization, team or individual. This is connected to a lack of adaptability: Mental flexibility, speed and agility are reduced, mental blockades or organizational blockades take up more and more emotional space. These blockades in the system (organization, team or individual) form in a systemic sense, and interact with the environment (customer and market, organization and team).

The task of transformation management is to renew this adaptation capability by suitable interventions, i.e. to remove blockades and to establish new behavior that is adapted to the changed environment.

In the context of interventions, be it with people, teams or organizations, the terms "change management", "transition management" or "transformation management" are used. Very often these terms are used synonymously. We use these three terms for different types of interventions. Figure 4.21 illustrates the characteristics of these different types:

We speak of change management when the current state (old), the desired new state (new) and the ways to them are known. This is a classic management task. The manager, a line manager or a project leader, knows the current and the future situation. The individual activities of the way from current to new are also known. Very often such change management processes are subsumed into a "rigid" system, which needs to be executed. We refer to the standard work of Doppler and Lauterburg [28] for a comprehensive presentation of change management with certain aspects of transition and transformation management.

[12] Here, we adhere to Luhmann, who regards communicative entities as system elements that generate a system, which is independent of the people who trigger this communication.

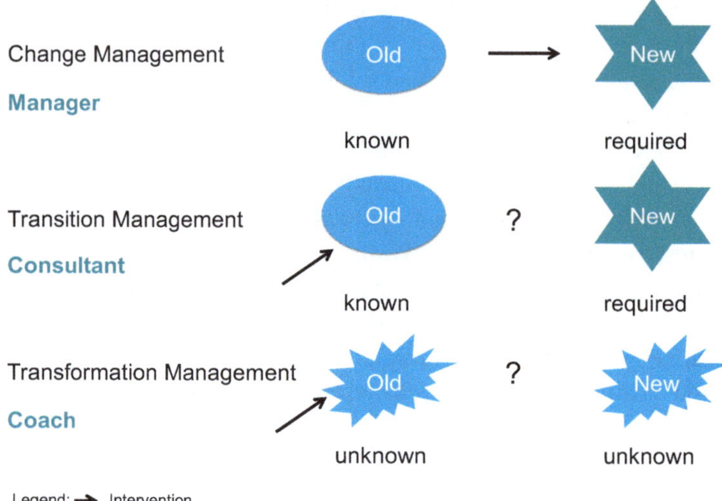

Fig. 4.21 Types of intervention

We speak of transition management, when both the old and the new situation are (more or less) known or are defined, but the path between them is left open. This type of intervention provides a certain framework, e.g. by defining targets or by settings with respect to vision and mission. The route from old to new is made more concrete. In this case, management acts like a consultant who intervenes in the current system to induce a transition to the new transformation management is characterized by the fact that a person, team or organization is not given any guidelines. "Old", "new" and "the way between them" are "illuminated" by the respective system itself. Transformation management is therefore essentially based on self-organization, and transformation management can only be acted on via coaching: The coach intervenes in the complex system, but does not know where the interventions will lead the development, and accepts the uncertainty associated with the interventions.

Both "old" and "new" are each characterized by a Dilts Pyramid and, of course, any time between "old" and "new" can be represented by an intermediate state using the Dilts Pyramid. The set of Dilts Pyramid from "old" to "new" form a state space, which is spanned by the transition.

It is irrelevant whether we are speaking of individuals, an organization, a network of organizations, or one or more individuals within a network, when dealing with transformation work.[13] In all cases, the states can be described by the characteristics of transformation work on the Dilts Pyramid. Please note that the logical level "identity" is the least

[13] In the following section we will only speak of transformation work, including all three types of intervention.

changeable. We describe this level through the temperament model and the model of basic needs. The four basic needs are connected directly to the four v-memes (red, blue, orange and green). The level of identity has a key function in the Dilts Pyramid, either looking downwards from this level or looking upwards from it. On the one hand, through the awareness of one's own identity, new values and basic assumptions, as well as abilities and behaviors are developed. On the other hand, a deliberate discussion and acceptance of the belongingness of one's own identity to a group, as well as interconnections with one's past and future with respect to vision and mission emerges. As we have already explained, this corresponds to the creation of sense.

Each of these manifestations of transformation work has its justification, whereby only transformation management results in being unrestricted due to intrinsic motivation of the participants. The term management is actually wrong in this case, since coaching is always a form of obstetrics. NLP uses "contentless" strategies, so-called formats, which form a framework for open and unbiased interventions through open questions and hints [29].

Change, transition and transformation management, can in principle, be based on two basic categories of leadership or interventions. Each system is determined on the one hand, by its micro-tier, and on the other, by its macro-tier (see Fig. 4.22). The social micro-tier is one of the agents' behaviors. The social macro-tier is the overall behavior of the system, resulting from the behavior of individual agents. Intervention could be essentially made on one of these two tiers. The replacement or transformation of individuals is an intervention on the social micro-tier and can have a massive effect on the macro-tier. In this way, it is very popular to replace project managers, managing directors or executive board members to induce change. In many cases, (perceived) inadequate leadership abilities of leaders are the motivation for this. In only a few cases, we need to be aware that the change is largely due to systemic interaction between management and organization. The replacement of one or more key agents is often the final measure to initiate organizational transformation.

Individual transformation work is an appreciative alternative, because, as we have seen elsewhere, transformation becomes necessary when behavior acquired through experience

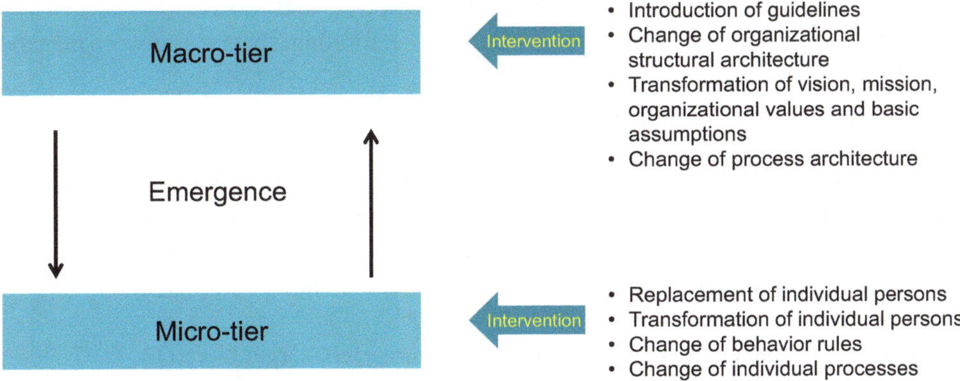

Fig. 4.22 Categories of intervention

(in the organization) no longer fits the new environment of the organization. Superficially considered, the transformation of behavior can be acquired through appropriate skill training. If carried out in this way, very often a change in behavior is not sustainable. Individual behavior transformation is only sustained by corresponding transformation on the upper levels of the Dilts Pyramid: For example, if behavior transformation is redesigned by individual transformation of basic assumptions and/or values and these are linked to new behavior patterns.

While interventions on the micro-tier focus on a system's agents (elements) and behavior, interventions on the macro-tier regulate the macro-structure of a system. As we have seen in previous chapters, there are three types of parameters that influence the occurrence of value-creating macro-structures and their manifestations:

- Setting parameters,
- Control parameters,
- Order parameters.

Value-destroying system structures arise when these parameters are not shaped "correctly". Thus, transformation work on the macro-tier is based on changing these parameters. So what are the parameters that must be defined for projects on the macro-tier? From the chapter on self-organization, we can name the following parameters.

The setting parameters in individual projects include, for example

- Project organization and rituals,
- Team composition,
- Local and temporal conditions (including distributed teams, locations).

The control parameters in individual projects include, for example,

- Limitation of the workload (also called work-in-progress in Agile Management),
- Appreciation of individual needs and associated motives and values,
- Strength-oriented assignment of team members, i.e. forming a balance between demand and skills,
- Forming a common identity and belongingness.

The transition between setting parameters and control parameters is partly fluid. Or in other words, setting parameters determine fixed conditions. If a setting is changed more or less smoothly, the setting parameters become control parameters.

Order parameters in individual projects include, for example,

- Collective vision and mission, common big picture of the target hierarchy,
- Focus on the project status via the big picture,
- Guidelines with common values and basic assumptions (in line with those from multi project management, see below).

Order parameters are therefore used to create a self-referential focus for the project team, therefore "aligning" the mental models of all team members.

The formation of macro-structures always emerges via self-organization. If the various parameters, as described in the section "Regulation of Complexity Through Targeted Interconnectedness and Self-Organization", are not adjusted along with each other, no value-creating macro-structure will develop.

A set of interacting agents does not necessarily form a system with a value-creating macro-structure. The goal of any intervention on the macro- or micro-tier is to create a macro-structure that does not show blockades and can adapt to new environments.

Organizations consist of a hierarchy of alternating micro- and macro-structures. Projects as system elements form a micro-tier. As a result of the interaction of the projects, a new macro-tier emerges that requires its own system parameters:

The setting parameters in multi project organizations include, for example,

- Elements and structures of multi project organization (i.e. Project Management Office),
- Decision-making bodies,
- Dependencies (i.e. between teams, other resources).

The control parameters in multi project organizations include, for example,

- Limiting the organizational workload (organizational work-in-progress),
- Connection to the company strategy, clear priorities,
- Joint resource management, taking into account availability and competence.

Also here, it is true that the transition between setting parameters and control parameters is partially fluid. If the tasks of the PMO are changed more frequently, this may act as a destabilizing control parameter.

The order parameters in multi project organizations include, for example,

- Collective company vision and mission, common big picture of the "enterprise" system,
- Emerging focus based on the big picture of the multi project management status,
- Guidelines for common values and basic assumptions.

Here again, it is true that the value-creating complexity of multi project management organization only occurs by self-organization. Now the question arises as to what should be done, if a new environment leads to blockades and thus a transformation becomes necessary.

Family therapist Virginia Satir [30] was probably one of the first to deal with the phases of transformation work. She identifies the following phases (see Table 4.5), which are still used as a basis for known phase models on transformation work:

Plotting the performance (understood as the ability to use the new environment in a value-creating manner) of a person, of a team or an organization over the six phases,

Table 4.5 Phases of transformation work according to Satir

Phases	Phase	Characteristic feelings
1	"Old" stable state, status quo, the new or the need for change arises	Irritation, pain, imbalance
2	The new or the need for change is articulated	Uncertainty: Foreign element penetrates into the system
3	Chaos, movement from a dysfunctional state to a functional state, from the familiar to the unknown	Anxiety, Fear
4	Integration of the new or of new learning experiences	New safety, new feeling of well-being, new hope
5	Stabilization of integration, phase of practice: Strengthening the new state	Maturation of the feeling of safety
6	New stable state, new status quo	Equivalence, harmony, wholeness, balance, open to new opportunities, wellbeing

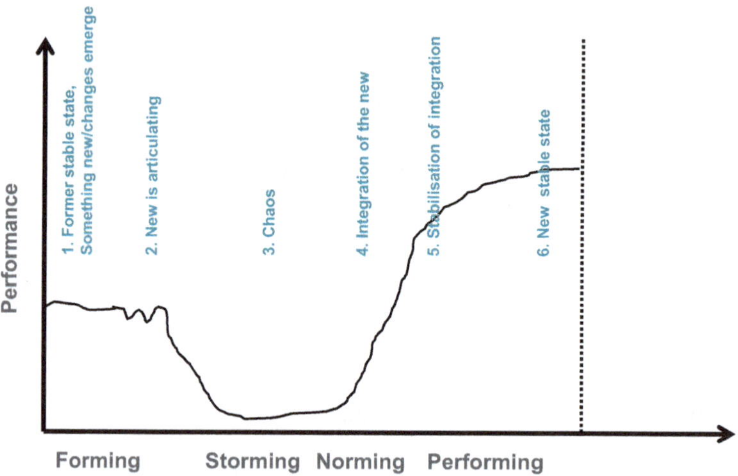

Fig. 4.23 Phases of transformation and performance

according to Fig. 4.23, a diagram results, which is widely equivalent to the diagram of team performance of the Tuckman phases. The phases of team development are included in the figure as reference points. The fact that the phases of team development follow the phases of transformation is easy to understand, because team development process is transformation work:

In the Tuckman phase "Forming", team members come together with their personality and current experience and background. We assume that prior to this, each team member is in a stable state. From the picture we can deduce that turbulence or chaotic conditions will occur soon if one of the team members is out of balance. After a period of careful mutual scanning, differences in interests or personality will be perceived. If the differences are large enough, they will also be expressed by one of the team members. In most cases, an attempt is made to compensate the differences, so that there will be fewer fluctuations in a team's mood and therefore also in team performance.

Only when differences, or the new, in the form of a new task, are perceived as too large, will team performance break down. The team has then entered into the second Tuckman phase "Storming", or the third phase of transformation according to Satir, "Chaos". This is where greater emotional contention with other team members and/or the task takes place. This phase is absolutely necessary, because without this emotional contention, no integration or stabilization can take place in the Tuckman phase "Norming" or phases 4 and 5 of the transformation, which are associated with a higher level of team performance. In order to achieve a higher level of performance, it is necessary to combine the new with the in place experiences in such a way that the old and the new are not juxtaposed, but where mentally, something completely new is created by integration. If the integration has stabilized through continued practice, a new stable state will have been achieved. Stable states tend to stiffen, i.e. they lose their ability to adapt to new environments. At a later time it will be necessary to re-enter the process of transformation. A new four-step transformation process then begins.

In the following, we assume that any transformation work follows the above pattern. On this basis we propose a transformation model consisting of four main phases, analogous to the Tuckman phases and in accordance with the Satir phases of transformation. The main phases can be shaped differently depending on the environment. We have connected the main phases with the central ideas of self-organization, namely setting, control and order parameters, as well as the resulting macro-structure. We assume that a complex system, whether of individuals, teams or organizations, can only transform itself by self-organization.

Figure 4.24 shows the phase model of transformation; according to the PDCA-cycle, however as a cycle, which has to be continually carried out with the changing environment.

Necessity for transformation is perceived in the "illuminate pressure points" phase. We call this "illuminating pressure points", because it is necessary that those who are to undergo a transformation perceive and feel the need for transformation, if necessary also in the form of pain. This necessity for transformation is, however, only perceived if individuals leave the pure behavioral levels of the Dilts Pyramid "to climb" to higher levels, and in an ideal case scenario, even step out from the Dilts Pyramid and look at the system from the outside. Only when this happens, can the need for transformation be accepted and articulated. It can also be said that a meta-perceptual position has been taken.

If there is willingness for transformation, a phase of disturbance and sometimes turbulence or chaos begins. In this phase, consciously or unconsciously, the new parameters of a future order, i.e. a new macro-structure, are searched for. This is the crucial phase of the entire transformation. The selection of parameters determines the success or failure of a transformation. The parameters searched for are setting, control and order parameters. Very often, parameters are chosen at the micro-tier, i.e. detail levels, which are regarded as

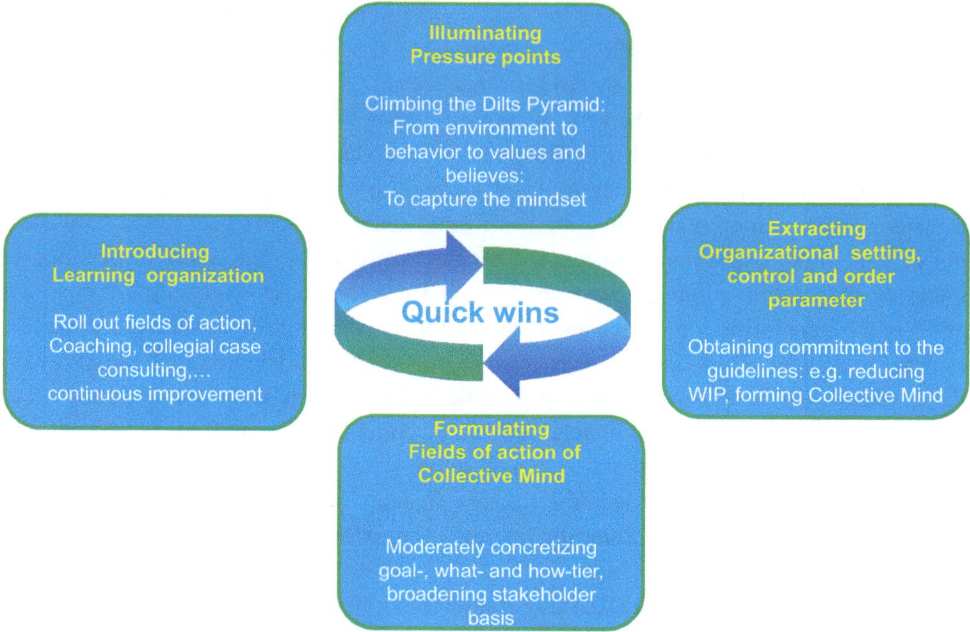

Fig. 4.24 Phase model of transformation

important: For example, new roll definitions or new processes. In many cases, however, these parameters are by no means crucial for the transformation, but belong to the third phase of the formulation of fields of action.

The formulation of fields of action is necessary to enable those involved to anchor to the system parameters according to their individual preferences.[14] When looking for the parameters, characteristics have to be taken into account: The transformation itself is subject to a Collective Mind process, i.e., it is necessary to ensure that the view to the goal is a collective view. For the process of the transformation itself, setting, control and order parameters are necessary, which need not be identical to those of the future new macro-state. The following examples will illustrate this.

In phase 4, "Learning Organization", the setting, control and order parameters are anchored in the organization, but arrangements are also made so that the organization is able to adapt more quickly and flexibly to future new environments.

The learning capacity of the organization will be sustainably improved by measures such as coaching and collegial case consulting. It is the "preparation" for a new transformation process. Since transformation work will not recognized by all those involved in a similar way, it is useful to support the entire transformation process with quick wins.

[14] Individuals with the temperament preference "S", need clear consequences for their fields of action, derived from the system parameters.

Quick-wins indirectly take on the role of control parameters for the transformation process: The motivation for transformation is ensured by means of quick wins.

In the following sections, we outline transformation examples or individual aspects of a transformation:

- The example “Bottom-Up Transformation” describes a transformation that has evolved from the requirements of the project managers of a company.
- The example “Top-Down Transformation” describes a transformation initiated by the top management of a company.
- The example “Coaching” outlines a coaching process for a project manager, who faces a new challenge.
- The example “Collegial Case Consulting” outlines the course of a transformation process, which supports the transition to a learning organization.

4.9.1 Example "Bottom-Up" Transformation

We describe the example as follows:

Initial Situation A software company generates sales by selling software and related customizing projects. The project management (PM) business processes are to be revised and this revision is carried out with external support in a core team of eight experienced team leaders and project managers. The PM business processes were introduced, but even the eight team members who have developed these new business processes do not apply the PM processes or do so only partially.

Analysis of the Situation There is no introduction of the new business processes. The project managers, who have co-developed the PM processes and are to apply these processes, cannot be part of the new processes with their current work environment. The project managers of the core team confirm that these new processes are very well suited to PM requirements. The introduction of the new PM business processes had so far (erroneously) been characterized as a change process rather than a transformation process.

Solution Outline A transformation process is initiated, with the coarse goal that at the end the project managers adopt the PM processes. The basic assumptions for the transformation process are that the project managers are prevented from implementing the PM processes by current setting, control and order parameters. The task is therefore to find the current parameters and generate new ones that trigger the transformation. This is to be carried out in such a way that during the transformation process a Collective Mind develops among project leaders, which will support the introduction of the new macro-state.

Approach Organizational transformation is carried out according to Fig. 4.24: The first two phases are carried out within the core team of project leaders. The following two phases are applied to the entire organization. Core team transformation begins on the micro-tier, i.e. the behavioral level of the team members. Since the team members are a good representation of all other project leaders, it is to be assumed that with the help of team members, current parameters will be found and new representative parameters generated.

Several workshops, set in a relaxed atmosphere, are scheduled, in which the phases "illuminating pressure points" and "extracting organizational setting, control and order parameters" are carried out in a fun atmosphere. The relevant fields of action are coarsely identified, but not formulated.

This procedure is carried out in a light-hearted way according to the format "10 steps of transformation". This format is based on the NLP formats "Six-Step Reframing" and "Meta-perceptual Positions", taking into account the Dilts Pyramid with its underlying basic assumptions and values [31]. Stories from the project business day are re-enacted and mentally processed. The format is carried out under the guidance of a team coach:

Step 1: A sponsor tells a story from a project that they themselves felt unsatisfactory and they subsequently want to change their behavior.

Step 2: The members of the core team each select a role to play from as a protagonist in the story.

Step 3: The actors re-enact the story. The sponsor observes and reports their perceptions and feelings. The other actors then relate their perceptions and feelings too.

Step 4: It is assumed that the sponsor always had positive intentions for their behavior. Positive intentions are separated from displayed behavior: For intention and behavior, values and basic assumptions are identified. Thus the shown intention is the bridge between the old and the new world.

Step 5: The core team looks for new behavior that suits the intention. The Dilts Pyramid is used here to illuminate the interaction of logical levels and, in particular, to analyze values and basic assumptions and adapt them, if necessary: Ways of behavior are analyzed by setting them in relation to the values and basic assumptions which determine them. The values and basic assumptions discovered, are checked for their validity in this environment and, if necessary, replaced by more suitable ones.

Step 6: The sponsor takes a look into the future: Are there any resistances to be expected that may hinder future new behavior? In NLP, this step is referred to as "Future Pace". Behavior is processed in the core team until the sponsor no longer sees any resistance in the future or if measures for solving theses future resistances have been found. Typical resistances are e.g. "I do not feel well," "My boss will not support this", "The customer will not participate".

Step 7: If, subsequently, a concept for new behavior exists, potential new capabilities associated with this new behavior, need to be identified.

Step 8: On the basis of the new behavior, the actors re-enact the story again.

Step 9: The sponsor observes and reports on their insights and feelings. The other actors then report on their perceptions and feelings. Once again, "Future Pace" and potential changes are added.

Step 10: The new behavior, new values and basic assumptions, capabilities, explanation of the old behavior, as well as suggestions for further adaption, are recorded in written form. The first fields of action are identified, on which the new behavior has an impact. If possible, the first clues on the adaption of the fields of action are included.

An example story, or situation spotlight from a story, could be that of a project manager not taking sufficient time to prepare an acceptance date with the customer and therefore anticipating possible alternative actions for the date with customer. The customer exploits the situation, and refuses to accept. In retrospect, it turns out that with sufficient preparation there would have been no reason to refusal to accept.

This Results in the Following Parameters for the Initial Situation

The project manager's positive intention is combined with the values "flexibility" and "customer orientation": With his behavior, the project manager tries to satisfy as many of his projects and clients as possible, with the result that there is not enough time for each project. Professionalism decreases. In the meeting with the customer confrontation is avoided: Positive intention is again, customer orientation, of course, accompanied by the fear of annoying or even losing the customer. The high number of projects running parallel is identified as a control parameter which is "set" as value-destroying.

A basic assumption is worked out: "The person who has as many parallel activities as possible is a good employee or project manager". Flexibility and customer orientation are identified as values. Basic assumptions and values are identified as the order parameters of the behavior.

As fields of action, the new design of customer management and o time management are identified.

This Results in the Following Parameters for the Newly Re-enacted Situation

The positive intentions represented by the values "flexibility" and "customer orientation" are retained. A new basic assumption is formulated: "A good project manager is one who takes their time to prepare their appointments". These values and this basic assumption are incorporated as order parameters.

New behavior is formulated and "tested" by re-enacting: The customer is guided by proactive action. In particular, "rules" for regular contact are defined as rituals.

These new ways of behavior, basic assumptions and, where applicable, values, are added to the fields of action. As a control parameter, a benchmark for the number of projects in parallel per project leader is also set. In the action field time management, this results in further detailing by marshalling rules for the organization of the diaries.

Based on about a dozen stories and their "transformation", all system parameters are identified: It turns out that in the stories, again and again, the same setting, control, and order parameters appear in different guises.

With the elaboration of all system parameters and a Collective Mind for these system parameters, the transformation phases "illuminating pressure points" and "extraction of organizational setting, control and order parameters" are completed.

The fixing of the "new" on a broad organizational basis begins with transformation phase 3, "Formulating fields of action of the Collective Mind". The core team is requested to concretize the collective understanding in the fields of action.

In principle, the system parameters identified in the previous phases would be sufficient to change the system "company". Experience has shown, however, that a transfer from the few system parameters into the daily world of action, is not or could not be made by all members of the organization. It is therefore necessary to derive concrete standards for action. Sometimes the term 'concrete specifications for practice' is used. This is due to the fact that more abstract guidelines are often unable to be combined with everyday activities. For example, if the control parameter "number of tasks" is formulated in the sentence "ideally, the number of concurrent tasks to be carried out should be one", this would ultimately not be interpreted according to its original intention for everyday life.

If interpreted appropriately, a respective employee would have to limit the number of meetings per day, schedule interspaces for pre- and post-processing of meetings, save certain periods of time for creative activities, and so on. And it is precisely this transfer performance, which is carried out when detailing the fields of action.

Together with the system parameters formulated as basic assumptions, the "what" and the "how" of the fields of action are added to a guideline document. This guideline document is the basis for anchoring transformation in the organization and is signed by the core team and management. At this point, a bottom-up approach is supported by a top-down approach. Signing the guideline document by management is an important symbolic act, which acts as an anchor and therefore constitutes an "independent" frame of reference. All employees now can refer *normatively* to this anchor.

Having done this, phase 3 of the transformation "Formulating fields of action of Collective Mind" is completed and the roll-out of the action fields begins. In this example, the roll-out begins with a five-day training course, in which the PM processes the guideline, and central models of the social techniques for complex systems are imparted. It is important that the concepts are all learned through practical exercises from the world of the project managers. The ratio of theory and practice is roughly 1: 5. This is necessary in order to stimulate mental transformation by practical exercises on actual project examples. Mental change here really means the construction of new neuronal connections. We emphasize that after training, project leaders need to first have the skills of self-reflection and thinking in systemic relations. According to the insight that old ways of behavior are only hampered by new ways of behavior, it is necessary to develop new ways of behavior through further learning. This is done by coaching and collegial case consulting. In the following sections, we will discuss these examples in more detail.

Finally, we would like to emphasize a generally valid conclusion, which is connected to the insight "neural structures of old ways of behavior don't vanish completely, but remain inhibited by new neural structures of new ways of behavior": If the transformation process on the organizational macro-tier is not fast enough, employees of the organizations will be found either partially or completely following old organizational ways of behavior. Associated with this, is the fact that they are thrown back into their own appropriate attitudes. The new ways of behavior, which are painful, and were illustrated in training, coaching, or in collegial case consulting, do not receive any support and are therefore unable to inhibit the old ways of behavior. The transformation process begins to fail. The factor time to the establishment of new organizational macro-structures is therefore critical for the success of a transformation. It is therefore important to keep the structures associated with the transformation process as stable as possible for a few months, so that the old ways of behavior are inhibited. For this reason, it is necessary that there are a specific number of promoters to continually promote the guidelines of the transformation, and that top management maintains correspondingly rigorous adherence.

4.9.2 Example "Top-Down" Transformation

We will now describe an example of a "top-down" transformation:

Initial Situation An e-commerce company is extremely successful and generates a billion revenue with a team in the lower three-digit range. In order to meet increasing demands, two-day project management trainings take place. Within the context of these trainings, the collegial case consulting method is used to mentally combine the newly acquired skills with experience gained from current projects.

Analysis of the Situation It turns out that the company has a high level of value-destroying organizational complexity in its processes and structures, which has barely been recognized from outside the company. Management can be convinced that the negative effects will become visible to customers in the medium term and that there is a subsequent need for action.

Solution Outline Introduction of a transformation process accompanied by external coaches, with the goal of making the organization more efficient and effective by taking a few simple measures. It is therefore all about the implementation of new setting, control and order parameters, which meet this goal.

Approach For this purpose, an extended management team, consisting of eight members, passes through the "illuminating pressure points" phase. Based on visualized business process architecture, the pressure points are marked and named using Post-it notes. In the context of "story-telling" with subsequent discussions, the pressure points are illuminated and the basis for a Collective Mind arises for carrying out the following necessary steps. Simultaneously, a certain amount of confusion is noticeable, as the question arises as to whether, and how, the number of pressure points could be resolved by just a small number of measures. From the desire to extract only a few key measures, an opportunity arises for the coaches to now offer a few principles as a guideline and that this will be finally accepted.

Five central principles are formulated, each consisting of only one sentence. The guideline itself represents the order parameter. The individual principles in the guideline act as control and setting parameters. A corresponding action field is defined for each principle. The most important organizational control parameter, as in the previous example, is the limitation of the number of projects being processed in parallel. Therefore, a new multi project management structure will be set up consisting of a decision-making body and decision-making principles. This control parameter is flanked by three further setting parameters: The first parameter is used to build up a uniform understanding of the role of the project manager as a leader and strengthen their position within the company. With regard to the project manager , this organizational setting parameter even acts as identity-building and thus as an individual order parameter. The second parameter focuses on the strengthening of systemic thinking and acting, because it is necessary for the project team to be perceived as competent in systemic thinking and acting and to be supported by structural measures such as the implementation of business and IT architects. And the third setting parameter provides for the structured integration of new projects following a well-defined initiation process into the multi project management structure.

After the design of the guideline, phase 2 of the transformation process according to Fig. 4.24 has been completed. The phase "Formulating fields of action of Collective Mind" begins by bringing more employees into the transformation process. In the present example, only the guideline has been further concretized so that each employee can combine their field of experience with the newly required individual and organizational behavior. The phase "Introducing learning organization" was built up by various workshops for all employees of the company as well as by individual coaching and collegial case consulting for project managers.

4.9.3 Example Coaching

The individual coaching of project leaders should "always" include three components:

- Personal development and improvement of leadership and communication behavior,
- Theory and practice of tools for the management of single projects,
- Theory and practice of tools for multi project management.

With regard to the training in the use of tools, this type of coaching has as strong consulting part, where the coachee is advised on how to use the PM tools in the context of the respective project. The social techniques tools presented in this book are, on the one hand, the "armamentarium" of the coach, but can also be used to train the coachee in applying them.

The current example is part of the PM coaching component "personal development and improvement of leadership and communication behavior":

Let us assume that Ms. Snyder has become the project leader of a complex and innovative project. After the first project team meetings, Ms. Snyder states that project management of this project is very difficult for her, although she has successfully mastered various projects in the past. She asks her supervisor for advice. The supervisor provides her with a personal coach. Because of the characteristics of the project, the coach arrives at the following analysis for Ms. Snyder's v-meme structure: She wants to implement the project successfully, but not under all circumstances (v-meme-level 3, red is not highly developed). Respect for what her company has achieved so far is huge and she is afraid to initiate new things for fear of a breach with the old (v-meme-level 4, blue is highly developed).

It is extremely important for her to work in a team and to treat others fairly, and in return, she likes to be treated fairly as well (v-meme-level 5, green is very strong). Since the present project is complex and innovative however, entrepreneurial spirit and insight into the relativity of existing systems are demanded from her. Therefore, transformation work is required, which emphasizes the v-meme-level 4 orange more and links it to the other v-memes. This ability for linking is expressed by the v-meme-level 6 yellow.

The following figure shows Ms. Snyder's v-meme profile according to the Spiral Dynamics model (see Appendix "Fundamentals Spiral Dynamics") prior to beginning the coaching process and the intended v-meme profile after the coaching process (Fig. 4.25).

The size of the colored ellipses in the figure symbolizes the strength of the v-mems perceived by the coach (the first two levels and the last level of the model were not displayed, as we do not address them in the coaching process). The task of the coach is, in particular, to significantly expand Ms. Snyder's entrepreneurial spirit (level 5 orange), by applying interventions and, at the same time, to reduce her orientation with the status quo (level 4 blue). For this, it is necessary to strengthen her assertiveness (level 3 red) and stimulate her ability to "think in interrelationships" (level 7 yellow). This is to be carried

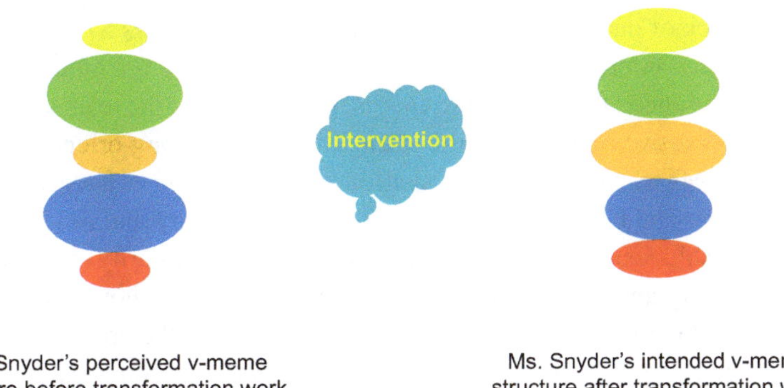

Fig. 4.25 Ms. Snyder's v-meme profile according to the Spiral Dynamics model

out in such a way that her self-confidence with respect to her strengths increases and she recognizes the importance of the interactions of all members in the projects, understands them and finally also learns to influence them in a well-directed way with respect to the project.

The coach decides to use four tools: The MBTI Temperament model, the motive and value model and Spiral Dynamics and the Dilts Pyramid. (In coaching, these four models come together in an integral form, see section "Neuroleadership" and the appendixes "Fundamentals Consistency Theory", "Fundamentals Reiss Motive Profile" and "Fundamentals Dilts Pyramid").

In the first block of interventions, the coachee learns to recognize her own temperament and associated manifestations in her communication. Learning of the MBTI model is supported by appropriate exercises for team analysis and stakeholder analysis. In this way, initial patterns of communication in the team and with the stakeholders are recognized, analyzed with specific examples, and initial alternative behavior rehearsed.

Communication and leadership competences will therefore be greatly augmented (usually). A further step in this direction, and an enhancement of the appreciation of the variety in the team, is made by the introduction of the motive and value model. The leadership model, as described in the previous section will be used for concrete leadership situations: Concrete leadership situations are "gone through" according to the models outlined above. On this basis, hypotheses are formulated for the leadership of the respective guided persons. These hypotheses are examined in practice by Ms. Snyder. In the following coaching session, the effectiveness and impact of the hypotheses are analyzed and established as appropriated corrective actions for practical application. The coach then guides the coachee to carry out the PDCA cycle and consciously form it. Over the course of 2–3 weeks and with about 2–3 coaching lessons per week, Ms. Snyder experiences a series of flashes of inspiration. (At the beginning of the coaching process, it may be useful to teach the basics of the models in a 4-h session. Later on, coaching sessions can take place over the telephone.) Of course, between coaching sessions Ms. Snyder is asked to carry out various "homework tasks" in which she applies what she learned in practice according to the PDCA model.

In the first block of intervention, Ms. Snyder assessed herself as SJ (Guardian) with the values of safety and orderliness (v-meme level 4 blue) and team orientation (v-meme level 6 green) and experienced their related strengths. An analysis of the team structure revealed that she has at least two members in the team who have a strong orientation towards the orange v-meme. Ms. Snyder has stated that one of these two people always brings new ideas into the team and both are willing to take more flexibility and risk. She now understands her leadership task in such a way that she should (temporarily) be able to reject her preference for safety and orderliness in favor of the project and should be able to succeed in using the interplay of different personalities with respect to the project. This also means that, when communicating, she should be able to perceive synchronously the current active motives and values of her behavior, reduce her associated preferences at suitable moments, and actively bring them in at other, more suitable moments.

The second block of interventions is based on the first block. The interplay of personalities in the team is not without problems. It is always important to recognize blockades and dissolve them. Ms. Snyder has learned to perceive patterns of behavior in herself and in others too. However, she does not yet know how to resolve these blockades. The models of Spiral Dynamics and the Dilts Pyramid are used here. The Dilts Pyramid serves as a "master model": The environment in which a blockade occurred in the team is illuminated. For this purpose, further additional models can be used e.g. the diamond model for illuminating the basic assumptions of the team members associated with the project type dimensions. The respective behavior of the team members is analyzed in detail. If necessary, biases which lead to further underlying basic assumptions are revealed. These basic assumptions are viewed in the light of temperament, motives and values, and this in turn helps to understand these basic assumptions even better and, in addition, to treat them with appreciation. This block of interventions by the coach is thus used to assist the coachee in recognizing and processing systemic connections in a professional manner. We speak of professional processing, because the coach teaches the coachee to gradually apply the initially used coaching tools and therefore implement an intervention based on social techniques. In particular, the Dilts Pyramid and Spiral Dynamics help to develop a linked, more systemic view, thus developing the v-meme level 7 yellow.

4.9.4 Example Collegial Case Consulting

In addition to coaching, collegial case consulting is a measure used to extend the already described process of personal development to team level. Collegial case consulting is a special form of experience exchange and of learning between colleagues, i.e. colleagues advise other colleagues. Colleagues can be from a project team or from an organizational department. In either case, collegial case consulting has a strong team-building and team-coaching dimension.

We have had very good experience with the following procedure:

Step 0: A colleague or supervisor (a colleague with a lot of experience or a coach) moderates the collegial case consulting.

Step 1: A colleague, the sponsor, relates a case (a story) to his colleagues. The story should be told very concretely (i.e. no generalizations, biases or deletions). The story should be visualized during its telling and should be limited to 3 min (maximum 5 min). The colleague closes his story by asking for advice from his colleagues.

Step 2: Colleagues ask questions within a time limit of 3 min (up to 5 min maximum) to help to understand the story better. Discussion or "judgmental clues" on other colleagues' questions are not allowed. The sponsor answers the questions; if necessary they can use visualization again.

Step 3: The colleagues offer propositions for solutions within a time limit of 3 min (maximum 5 min). Ideally, the solutions should be based on the models learned. The sponsor picks up all the solutions without making any comments, and writes down the solutions on a whiteboard or Flipchart. Discussion on or "judgmental clues" to the suggested solutions are not allowed.

Step 4: The sponsor informs the group which of the proposed solutions they want to take to try out in practice. They specify whether, and for which of the proposed solutions they see implementation problems. In some cases, they would have had already implemented proposals and they will describe what the impact of this implementation has been in the past. This step takes between 5–10 min, depending on the number of solutions.

Step 5: The solution proposals are discussed by colleagues. In many cases, further solutions arise, or the solutions proposed will be enhanced in a qualitative manner. The time frame for this is usually 30–45 min. During the discussion, personality preferences of the colleagues (temperament, motives and values as well as basic assumptions) will become visible. Since all colleagues have closely experienced the respective communication situations, the supervisor now has an opportunity to use these situations to learn and to combine the models with experience.

Step 6: The supervisor also provides a solution based on the learned models. The time frame is set to 15–30 min. This relatively large time frame arises from the fact that the models will be discussed intensively and applied in detail within the context of the proposed solution.

The entire time frame for collegial case consulting is usually 1.5 h.

Collegial case consulting is a wonderful tool for the sustainable development of the learning capabilities of an organization. All the tools of this book are used in professional solution finding. For example, the description of social interactions, rather than being based on meaningless descriptions of people, is instead, based on social network models, MBTI, and motives, values and basic assumptions. In this sense, a project characterization takes place by means of the diamond model. Based on these models, on the one hand a common collegial language is developed, and on the other hand, structure with a high level of variety is used.

The master model is the Dilts Pyramid. It is used to make mental blockades visible, which occur in the descriptions, in the suggested solutions of individual colleagues, or between colleagues. Figure 4.26 shows an example we have chosen, which refers to the dilemma representation of [32] by means of the Dilts Pyramid.

We could imagine that two colleagues have two different solutions for the improvement of a multi-project organization. One colleague may say: "First of all, we need to take full advantage of the resources in Project X, they are twiddling their thumbs and also carrying out some work twice." The other colleague says: "This is nonsense, the whole system of 'multi-project organization' is to utilize more efficiently, there are obviously too many dependencies between the projects, and it makes no sense to optimize locally."

A first step in reviving value-creating communication in such an example is, as shown in Fig. 4.26 to visualize and generalize the statements as provocative statements. In this process, the colleagues become quickly aware that both statements are based on models.

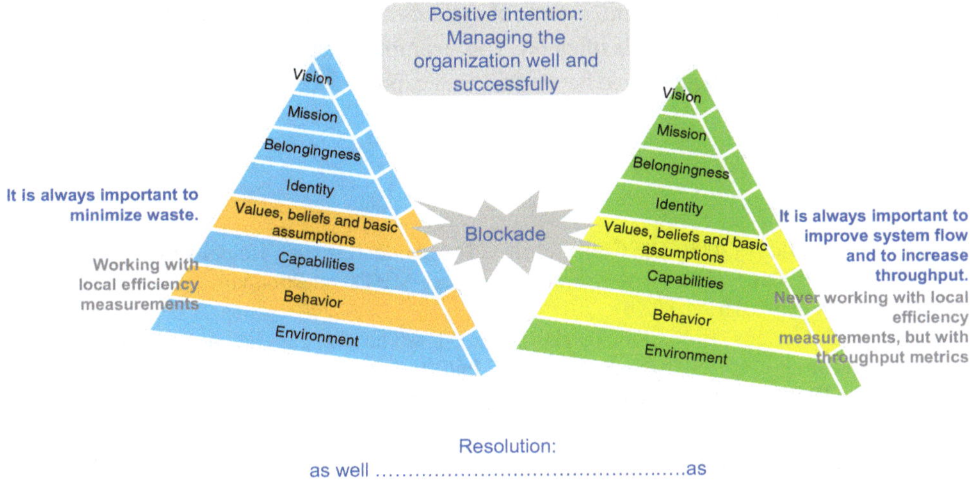

Fig. 4.26 Dilts Pyramid and mental blockades

These can be mental models that have been learned as rules of thumb during the course of practical experience, or they can also be explicit models, such as those taken from Lean Management or Critical Chain Management. Models are always mappings, and it would be fatal to give them an absolute truth. This is the next insight which occurs, and connected with this is the further insight that different points of views may definitely have the same positive intention. This leads to the crucial realization that it is quite useful to pursue both approaches. In many cases it is not an "either … or" but a "as well … as".

In our example, this means that both the local optimization, as well as system-oriented optimization will be pursued by means of throughput-indicators. In this example, we linked local optimization to Lean Management and system-oriented optimization (i.e. bottleneck-oriented) to the MPM Critical Chain Approach. In the chapter "Consequences for Management Systems" we will come back to the differences associated with these approaches.

It is important to note here that the two colleagues in our example recognize that both approaches are models (rules of thumb) and should not be seen as the absolute truth. In the MPM Critical Chain Approach, system elements, as well as the throughput-determining system element, are modeled as simple (machine-type) system elements. This does not take into account the fact that this system element can completely change its properties as a result of interaction with its environment: Local interactions can have a considerable influence on the formation of the throughput-determining system elements. Waste processing on the "team" system level of a project that is not the bottleneck is, from the critical chain view of a multi-project management (MPM) system, a local (wrong) measure. However, on closer examination this (wrong) measure could change the entire mesh of effects and therefore also the bottleneck perceived until then. More generally, this means that our practice consists of many different systems (human, team, departments, companies, networks of companies) at different levels of abstraction (system element, system, system of systems ...) and it would be fatal to select only one view – but that is what models often do. This example illustrates the fact that many views must be taken into consideration and be worked on, i.e. preference given to the "as well ... as".

Literature

1. Köhler J, Oswald A (2009) Die Collective Mind Methode. Springer, Heidelberg
2. Keirsey.com (2016) Keirsey temperament sorter. MBTI questionnaire. www.keirsey.com. Accessed 14 Feb 2016
3. Dilts RB, DeLozier J, Bacon Dilts D (2010) NLP II – The next generation: enriching the study oft he structure of subjective experience. Meta Publications, Capitola
4. McAdams DP (2001) The psychology of life stories. Rev Gen Psychol 5(2):100–122
5. Grawe K (2004) Neuropsychotherapie. Hogrefe, Göttingen
6. Beck H (2013) Biologie des Geistesblitzes: speed up your mind! Springer Spektrum, Berlin/Heidelberg. Kindle Version
7. Beck DE, Cowan CC (2007) Spiral dynamics: leadership, Werte und Wandel. J. Kamphausen Verlag & Distribution GmbH, Bielefeld
8. Peters T, Ghadiri A (2013) Neuroleadership – Grundlagen, Konzepte, Beispiele: Erkenntnisse der Neurowissenschaften für die Mitarbeiterführung. Springer Gabler, Wiesbaden
9. Edelman GM, Tononi G (2013) Consciousness: how matter becomes imagination. Penguin Press Science, Kindle Version
10. Wikipedia (2014) Flow (Psychologie). http://de.wikipedia.org/wiki/Flow_%28Psychologie%29. Accessed 5 Dec 2014
11. Duchmann C, Töpfer A (2013) Neuroleadership der Unternehmenskultur. Zeitschrift Führung + Organisation Schäfer-Poeschel 6:394–403, Stuttgart
12. Blackmore S (2005) Die Macht der Meme oder die Evolution von Kultur und Geist. Elsevier Spektrum Akademischer Verlag, Munich
13. Cameron K (2016) Organizational culture assessment instrument. http://www.ocai-online.com/. Accessed 14 Feb 2016
14. Schneider WE (1999) The reengineering alternative. A plan for making your current culture work. McGraw Hill Book
15. Oswald A, Köhler J (2013) Schnelles und langsames Denken in Projekten: Zur Beherrschung von Unsicherheit in Projekten, Teil 1. projektManagement aktuell 5:30–36
16. Oswald A, Köhler J (2014) Schnelles und langsames Denken in Projekten: Zur Beherrschung von Unsicherheit in Projekten, Teil 2. projektManagement aktuell 1:25–31
17. Kahneman D (2010) Schnelles Denken, Langsames Denken. Siedler Verlag, Munich
18. Cleese J (2012) John Cleese on Creativity. https://www.youtube.com/watch?v=Qby0ed4aVpo. Accessed 14 Feb 2016
19. Oswald A, Köhler J (2012) Mit dem Projektnavigator zum Projekterfolg. In: Lang M, Kammerer S, Amberg M (eds) Perfektes IT-Projektmanagement. Symposion Verlag, Dusseldorf, pp 167–192
20. Keirsey D (1984) Please understand me. Prometheus Nemesis Book Company, Del Mar
21. Bridges W (1998) Der Charakter von Organisationen. Hogrefe-Verlag, Göttingen
22. Schein EH (2004) Organizational culture and leadership. Jossey-Bass, San Francisco
23. Oswald A, Köhler J (2010) Wechselwirkende Organisationen, Teil 1. projektManagement aktuell 5:14–19
24. Oswald A, Köhler J (2011) Wechselwirkende Organisationen, Teil 2. projektManagement aktuell 1:36–41

25. Keirsey D (1998) Please understand me II. Prometheus Nemesis Book Company, Del Mar
26. Sackmann SA (2007) Assessment, evaluation, improvement: success through corporate culture. Bertelsmann Stiftung (eds) Verlag Bertelsmann Stiftung, Gütersloh
27. Hofstede G, Hofstede GJ (2005) Cultures and organizations. McGraw Hill, New York
28. Doppler K, Lauterburg C (2008) Change management, 12th edn. Campus Verlag GmbH
29. Giesen R, Kluczny JW (eds) (2011) Coachingperspektiven. Impulse für die Praxis. DVNLP e.V., Berlin
30. Satir V, Banmen J, Gerber J, Gomori M (2000) Das Satir-Modell: Familientherapie und ihre Erweiterung. Junfermann Verlag, Paderborn, Buchexzerpt erstellt von Martin Schütz Juni 2007
31. Mohl A (2010) Der große Zauberlehrling. Das NLP-Arbeitsbuch für Lernende und Anwender. Teil 1 und Teil 2. Junfermann Verlag
32. Techt U (2015) Projects that Flow: Projekte in kürzerer Zeit. ibidem, Kindle Version

5 Consequences for Management Systems

In the previous chapter, we presented various models of social techniques and illustrated these models by examples of their application. In this chapter, we would like to apply a few of these models to existing management systems, in order to better understand what they can and cannot do. We will also use "fluid organizations" to throw new light on to the "classic" dilemmas of project – line and virtual team. We are not aiming to cover these topics exhaustively using the above models – as that would be almost impossible. However, we do aim to do, is to explore new avenues in each subject using selected models and their application.

5.1 Mindsets: Lean, Agile, Critical Chain and Innovative

Within the context of the Lean and Agile Management Systems, Lean Mindsets and Agile Mindsets are often spoken of, without being clearly defined or understood. Earlier, we defined our understanding of a mindset as an attitude that someone shows in a particular context. We model this context-specific attitude by means of the Dilts Pyramid. In this respect, the Dilts Pyramid model can be applied to an individual, a group or an entire organization.

Table 5.1 shows four mindset examples. At this point we would like to emphasize that these are mindset manifestations that typically reflect the respective mindset according to the authors' understanding. By typical we mean that a respective mindset as described here, will rarely occur in a "pure form". It is more likely that a hybrid will occur, or a form that contains another additional typical mindset. "Classification" is conscious simplification, which helps recognize the essential in "the contrary".

Some readers may conclude that what is described either does not meet with their understanding or does so only to a limited extent. And this is exactly the reason for

A. Oswald et al., *Project Management at the Edge of Chaos*,
https://doi.org/10.1007/978-3-662-48261-2_5

Table 5.1 Examples of mindsets: Agile Scrum, Lean, Critical Chain MPM, Innovative

	Scrum Agile mindset.	Lean mindset	CC MPM mindset	Innovative mindset
Vision	Our working world is more human.	Our working world is better.	Our working world is optimal.	Our (working) world is new.
Mission	In the team, we solve every complex task with "empiricism".	We impress our customers with our products and at the same time increase business value.	With our theory of constraints, we define the optimum operating point for a system of projects.	We create the world of tomorrow with theory x or product x.
	Find solution and create effect.	**Increase efficiency and effectiveness.**	**Increase efficiency.**	**Create something new.**
Belonging-ness	We belong to those, who solve a complex task by delivering early and frequently.	We belong to those, who create efficient, customer-oriented products.	We belong to those, who understand a system of projects.	We belong to those, who push the world forward.
Identity	We are problem-solvers who respect their customers and expect respect from their customers.	We are entrepreneurs who create benefit for their customers and thus earn money.	We are entrepreneurs who optimally design a system of projects.	We are researchers. We are pioneers.
Values, belief systems	Values: Transparency, focus, courage, openness, commitment, respect. Principles: Early and frequent deliveries are possible thanks to iterative checking and adaptation. "Self-organized" teams make the working world human and create added value.	Values: Quality, customer orientation, reliability, standards. Principles: Waste must be avoided. The added value for the customer is decisive.	Values: System flow, transparency, reliability, stability. Principles: A system has only one bottleneck, which determines the flow of added value. High resource utilization is not a criterion for optimum flow.	Values: Creativity, curiosity, adventure. Principles: New is fun, new is interesting.

Capabilities	Agile Operational Framework Scrum: Work in a team, carry out an iterative operational framework with discipline, understand and implement the product vision, demonstrate solution competence as service provider.	Lean model- and method framework: Understand and continuously question own system of processes, see and implement added value for the customer.	Critical chain/ToC method framework: Be willing to understand the optimization of the added value of a project organization as the “definition” of a system at macro-tier.	Self-organization framework: Work in team, creativity, think out-of-the-box, identify trends, slip oneself into other contexts, shape the Collective Mind.
Behavior	Execute Scrum operational framework: Iterative search for a solution.	Execute Lean Operational Framework: Continuous optimization of the system.	Execute critical chain/ToC operational framework: Adjust system flow.	Execute self-organization: Iterative searching of Collective Mind, interconnect and develop.
Context	A complex task needs to be solved and the solution is assessed based on the principal/customer satisfaction.	A known product is to be created (Lean Production) or developed (Lean Development), with associated process improvements. Continuous improvement evolution is desired.	Under given underlying conditions, an existing project system is to be systemically optimized.	Something new is to be created. A possible “revolution” is desired.

textualizing a mindset. Along with the neurological levels, it helps make visible, implicit understandings held by different individuals or organizations, and where required, to set a dialogue systematically in motion.

We have selected the Agile Mindset of the Scrum Agile Operational Framework [1], the Lean Mindset [2, 3], the Critical Chain Multi-Project Management (MPM) mindset [4] and the manifestation of an Innovative Mindset [5].

In all four mindsets, it is common that a description of the topmost neurological levels, from vision to identity, is probably nowhere to be found. A description usually starts at the level of values, and underlying basic assumptions are also rarely explicitly expressed. What is always well described is the level of behavior via an operational framework or best practices.

By this comparison, it is easy to recognize that the described mindset characteristics have completely different roots. The Agile Operational Framework Scrum arose from the desire to establish a team-based working world based on individuals' needs. In the next section, therefore, we look at the extent to which Agile Management serves individuals' basic needs. This is also one of the roots of the Lean Mindset, however, the main root may be the thought of as increasing efficiency, or in other words the elimination of value-destroying complexity in the form of "waste". In order to decide what is "waste", a guideline is required, and this is customer value, or the value attached to the product from a customer perspective. The term "guideline" is only another word for an order parameter. The Critical Chain (Multi-) Project Management method mindset is based on the Theory of Constraints. Here, too, a form of "waste" is eliminated, namely the waste created at the system bottleneck. Whereas Lean Management is aware of and deals with a lot of different forms of "waste", the Critical Chain Method takes a systemic perspective and says that for the performance of the system, only the performance of the most influential bottleneck is important. The chosen manifestation of an Innovative Mindset is selected in accordance with self-organization and detailed by the Collective Mind Method.

Imagine now for a moment, four team members with these four different mindsets, form a project team. If the four team members are not able to make their own mindsets visible by means of meta-perceptions and mindfully perceive those of the others (control parameters), massive conflicts are inevitable. And yet, by disclosure of the mindsets one can immediately recognize that the system elements are completely different: For the Agile and Innovative Mindset, the system elements are individuals, in the Lean Mindset these are the processes and in the Critical Chain MPM these are the projects. It is also easy to see that different focusses exist. The Scrum Agile Mindset is about solving a complex task, but this is neither the focus of the Lean Mindset, nor the Critical Chain MPM Mindset. The Innovative Mindset is also concerned with solving complex questions, but those with their origin in the team itself. The Scrum team on the other hand, receives the task from outside, from the product owner.

All four mindsets contain elements of self-organization, i.e. there are setting parameters, control parameters and order parameters. Table 5.2 lists these system parameters, Bateson's stages of learning of the four mindsets and the related management systems:

Table 5.2 Setting, control and order parameters of management systems

	Scrum Agile Management *Find solution and create effect*	Lean management *Increase efficiency and effectiveness*	Critical Chain MPM *Increase efficiency*	CM Innovative management *Create something new*
Setting parameters	Shielding in space and time, Scrum operational framework, the appreciation the team receives from the environment.	Business added value process.	Project initiation conditions.	Shielding in space and time, the appreciation the team receives from the environment.
Control parameters	Limitation of the Work-in-Progress, iterations, appreciation in the team.	Waste or value-destroying complexity, continuous improvement.	Bottleneck, Work-in-Progress.	Convergent fluctuations between "open mode" and "closed mode", limitation of Work-in-Progress, appreciation in the team.
Order parameters	Product vision, Product backlog, Sprint backlog.	Business value/ customer benefit, Value stream.	System and project fever curve.	Collective Mind target hierarchy.
Stages of learning	Learning II: Retrospective, Review.	Learning II: Continuous improvement.	Learning II: Basic assumptions in the system are questioned.	For learning IV: Use of the Dilts Pyramid: Stages of learning in the system are only limited by that of the individuals involved.

Let us first consider the order parameter of the four management systems (see also Fig. 5.1): Scrum has an order parameter similar to the Collective Mind System, which is divided into three parts consisting of Product Vision, Product Backlog and Sprint Backlog. Since the Product Vision and the basic description of the product in the form of the Product Backlog is introduced mainly from the outside by the Product Owner, there is a high risk that the team will identify itself inadequately with the requirements, or misinterpret the requirements and only operate on the How-tier, i.e. the Sprint Backlog. As a result, self-organization would be possible in the sense of self-management, but not in the sense understood here. Setting and control parameters are "monitored" by a particular role, the Scrum Master.[1] This includes, of course, compliance with the Scrum operational framework, but also the formation of mutual appreciation. The functions "adjusting setting and control parameters" and "adjusting order parameters" are represented by two separate

[1] The Scrum Master is outside the "team" system. From a systemic point of view, it is important that the coach is not part of the coached system.

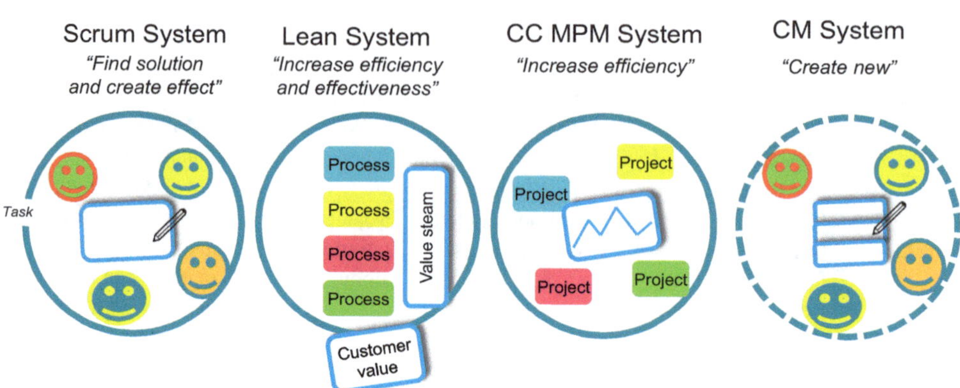

Fig. 5.1 The systems at a glance

roles, namely Scrum Master and Product Owner, which do not belong to the team which will form the self-organization. Therefore, the structure of the Scrum operational framework includes risks that may be an obstacle to self-organization, which will occur if the Scrum Master and the Product Owner are not aware of the principles of self-organization and do not subsequently apply them properly. Self-referentiality in the team is supported by specific forms of learning: This is the Retrospective, with its focus on the improvement of procedures within the operational framework, and the review, based on the feedback of the Product Owner and stakeholders on the delivered product increment. Learning III or IV is not explicitly supported, since the basic assumptions of the Scrum operational framework are not questioned.

Figure 5.1 shows the Scrum System with individuals as "system elements", and visualization via the Task Board, which supports in particular, the How-tier of the order parameter hierarchy, that is, task processing. The team is to a large extent, shielded in time and space from external complexity, only via the product owner. The conceptual formulation is entered into the system and verified at the time of review.

In this context, a frequently made experience from consulting practice needs to be emphasized: Those who introduce Scrum or other methods or operational frameworks, often tend to query various details: "When should specific sticky notes on the task board be moved, and to where?", "What is the Scrum master permitted and not permitted to do?" or "How long should a sprint be?" These detailed questions and their corresponding answers are completely irrelevant to the formation of self-organization or high level team performance. What is vital, is that these activities support the relevant system parameters.

We now come to the next system, the Lean System. It has its background in the desire to make production more efficient and to align the system to the production target, to a product with high value for the customer (while simultaneously minimizing manufacturer's costs). The system elements are therefore not primarily individuals here, but processes or process components. The assumed value, which the product has for the customer, is in this case, the order parameter. Ultimately, of course, alignment at this value is also carried out by individuals, individuals who have developed an idea of the potential value of the product. The order parameter "value" is broken down into value-parts, the value stream, along the added value chain. The value-destroying complexity "waste" has the function of a control parameter. As order parameters and control parameters do not "operate" directly on individuals, but on processes, self-organization can only emerge partially within the entire system.

The Critical Chain Multi-Project Management System (CC MPM System) has, as does the Lean System, not individuals as system elements, but projects. The aim of this management system is to adjust the throughput (flow) of projects through a system with given setting parameters to an optimum. "In contrast" to the Lean System, it does not search for different forms of waste in the entire system, but merely for the determining control parameter, the system bottleneck, and this is removed. If this bottleneck is removed, the next control parameter in the hierarchy of all of the control parameters, in other words, the next bottleneck, can be searched for and will be removed. This iterative approach is strongly oriented towards systemic efficiency. The flow of the system is the order parameter, which is visualized on so-called fever curves. Fever curves are 2-dimensional diagrams in which the project progress and buffer consumption are plotted time-dependently. Project progress corresponds to forecasted project progress and the buffer is extracted at the beginning of each project from the cost estimation. The learning supported by the CC MPM System remains at learning stage II, as the CC MPM System is not fundamentally challenged.

We have described the innovation-oriented Collective Mind System in detail in the section "Regulation of Complexity Through Targeted Interconnectedness and Self-Organization". Therefore, we will not give a description here, and merely refer to its high level of similarity to the Scrum System. Here, too, the system elements are individuals. Unlike in the Scrum System, system parameters are established by the team itself, as long as they do not violate the principles of self-organization.

In the sections "From Negative Words and Basic Assumptions" and "Philosophy of Complexity" we described the properties of a system by means of a group of individuals who are linked to each other without any connections, by a few connections based on rules, by many connections based on many rules, or by the "team orientation" value.

With our newly acquired understanding of complexity, we are now able to give the example "What is a system" a fresh appearance and classify the systems Scrum, Lean, CC MPM and Innovative (Collective Mind). Figure 5.2 shows the different variants of connections on the horizontal axis. These are special (discrete) manifestations of the control parameter "connections". "Energy" and "Capability to process information" are plotted on the vertical axis. "Energy" measures the extent to which the system can be moved as a whole by means of interventions. The "Capability to process information" measures the extent to which the system is able to respond adaptively to different interventions. The plotted values for energy and information processing serve merely as a rough qualitative orientation. Let us look at the two edge cases: If there are no connections between individuals, an intervention with respect to one individual (i.e. the individual is moved) does not cause a change to the other individuals: Therefore, quite a lot of "energy" is required

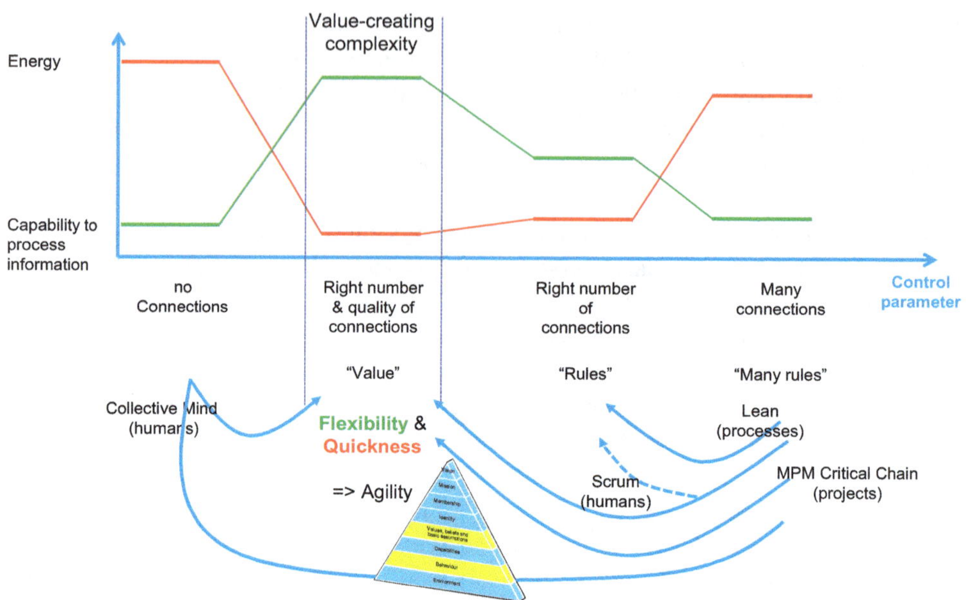

Fig. 5.2 Analysis of the example "What is a system?"

to move the system as a whole, because each individual needs to be moved separately. The "Capability to process information" is low, because the system as a whole does not respond. If a lot of connections exist, we have almost the same effect. These many connections make the system rigid. Therefore, a lot of energy is required to move the system and absorption of information is also very limited. The system is unable to create a representation or "internal model", of the intervention. If there are, however, a "right" number of connections, the system is able to respond to specific interventions adaptively: The few existing "right" connections enable the system to be moved as a whole by means of specific interventions. Ultimately, the internal rules of the system suit the intervention. We can once again illustrate the consequences of this statement for our example by assuming there are chairs between the individuals concerned. We now carry out an intervention and want to move the system as a whole. The chairs act as obstacles. The system cannot be moved as a whole, without breaking the existing rules, at least in some cases. Therefore the system cannot adaptively respond to the intervention.

Only if the rules are replaced by a collective understanding of a value, can the system behave adaptively: In our example, we have selected the value "team orientation" and in our example, we understand team orientation as such that a team will stay together, regardless of which form of movement is initiated by the intervention "a person is moved" throughout the system. In this case, possible obstacles are simply bypassed, because no rules are violated. Now, the system has a high level of adaptability.

Designing a system using values can, as we have already seen several times, result in value-creating or value-destroying complexity. Therefore, the selection of values for the adaptability of a social system is very important and implicitly, a portfolio of contexts is assumed for the selection of values. We associate the ability to create movement with minimal energy with speed, and the ability to respond adaptively to virtually any form of intervention with flexibility. According to our understanding, it is speed and flexibility together, that results in agility.

All four management systems (Scrum, Lean, CC MPM and Innovative) are focused on the area of value-creating complexity, i.e. the area of speed and flexibility. The basic idea of working "Lean", meaning to eliminate value-destroying complexity, is what all models, presented in this chapter, are based on. However, as we have seen above, these management systems operate on different system elements (people, processes and projects), and the means by which the objective shall be achieved, vary greatly. Lean reduces "waste", which means it eliminates connections that fail to offer (any longer) added value. Scrum tries to lock out external connections and create internal ones between the individuals who support their tasks. CC MPM eliminates value-destroying connections at the level of projects, by processing only as many projects in the system as the system is able to handle with the given resources. The innovative Collective Mind management system, in an ideal case scenario, assumes non-existing connections which, in the sense of the collective goal, still have to be established. In practice, however, individuals have a diversity of connections (contexts, experiences, assumptions and rules of thumb) which prevent the development of a Collective Mind, in more than just a few cases. In such cases, during the Collective Mind setup, transformation work with the team members' mindsets is necessary.

5.2 Agile Management via Scrum

In the previous section, we saw that Agile Management and the Agile Operational Framework Scrum are guided by the vision to make the working world more human. Expressed in terms of the models we have selected, this means that Scrum needs to be able to adequately take into account the four basic human needs. The question therefore arises as to what extent the measures and methods used in the operational framework can ensure this.

Table 5.3 shows the main measures and methods used in an operational framework.

For each of the measures or methods, the table shows which basic need, in our opinion, is specifically satisfied.

What is immediately noticeable are the small number of measures and methods. For this reason, the "simple operational framework" itself is a measure that satisfies the basic need for orientation and control: Through simplicity, value-destroying complexity is avoided and the feeling of straightforwardness and safety occurs. This is closely linked to the fact that decisions on the goal of the solution are outsourced in the form of product vision and product backlog. The product owner provides this information and the team looks for ways of implementation (separation of decisions). In particular, in the software development, where Scrum originally comes from, they desire clear requirements for development; a desire, which results from software developers' basic need for orientation and control.

Table 5.3 Scrum and the basic needs

Measure/method	Affection	Pleasure/prevention of pain	Orientation/control	Self-appreciation
Simple operational framework		Separation of decisions	Safety	
Face-to-face meeting	Team			
Exclusion of disturbances		Flow	Focus	
Daily stand-up			Ritual	
PDCA iterations		Satisfaction through results	Feedback	Self-efficacy
Sprint		Meaningfulness, success	Team commitment	Achievement of goals
Visualization	Common understanding		Common decisions	
Retrospective	Common understanding		Learning	Self-efficacy
Review			Feedback	

Working within a team, together at one location as far as possible, takes into account the desire for personal affection and is closely linked to the "exclusion of disturbances" measure during an iteration or sprint. Due to the fact that prior to an iteration there is a particular time slot in which the goal and tasks for a sprint are defined, and that with the beginning of the sprint no further disturbances are allowed, a high level of focus occurs during work and the team is able to achieve a team flow state.

Regular meetings and the disciplined implementation of a daily meeting becomes ritual and relieves the brain in its search for orientation.

The actions of the PDCA iteration and of the sprint are directly related to each other and have a very high impact on almost all basic needs: As the sprint goal is set by the team, this supports the need for control maximally. Because of this, the increased likelihood of achieving this goal contributes significantly in finding meaningfulness in the work. The team is strengthened in its co-operation, since it is the team as a whole, and not individuals, giving a commitment. All this contributes significantly to a feeling of self-efficacy.

Visualization of the joint activities via a task board generates self-referentiality, which generates the order parameter on the how-level (of the Collective Mind).

Retrospective and Review are two measures that support team learning after each sprint, on the one hand, with a view to the processes, and on the other, with a view to the outcome of the sprint (increment) and the associated consequences for the next iteration.

The measures and methods used in the Scrum operational framework have certainly been tried and tested many times, but ultimately for their selection, it is only important whether they support the leadership parameters for the implementation of self-organization. The interested reader is therefore left with the task of questioning the project management

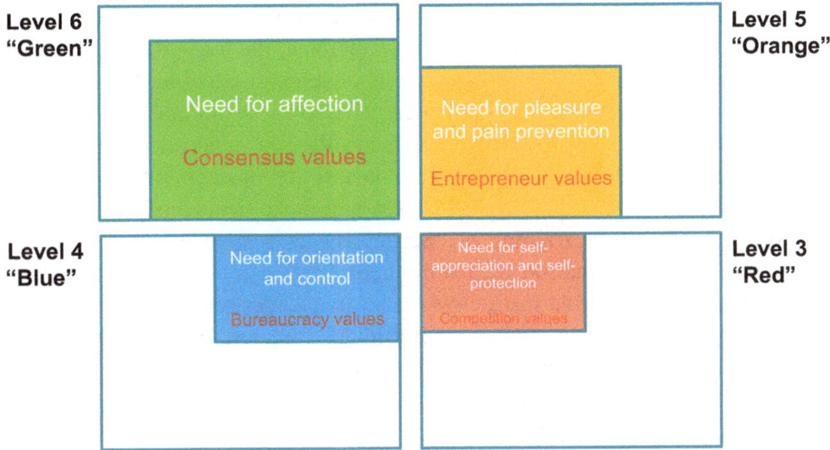

Fig. 5.3 Operational framework Scrum: Basic needs and v-meme (values)

methods used by their company in this respect. They can then ascertain whether the control parameters in their own project are set to support self-organization and subsequently agility.

As we have described in the section "Neuroleadership" and the appendix "Fundamentals Consistency Theory", the four basic needs are directly linked to the four v-memes level 3 to level 6 of Spiral Dynamics.

Figure 5.3 shows the strengths of the four v-memes of a typical Scrum team. A completely filled square corresponds to a maximum value. This figure is based qualitatively on a figure in [6], which is based on a survey of Scrum users on their Scrum culture.

Due to the typical v-mem structure of the Scrum operational framework obtained, it is possible to look at this in relation to the v-meme composition of the organization. The comparison gives direct insights into the acceptance of Scrum team culture in a respective company. And, of course, allows a comparative analysis of the team members' individual basic needs and the Scrum v-mem structure, to uncover any potential areas of friction between the Scrum v-mem structure and the basic needs and motives of individual team members.

In order to determine v-memes in the work environment, we refer the reader to the questionnaire by Peters and Ghadiri [7] mentioned in the section "Neuroleadership" or to similar culture assessment question catalogs [6, 8, 9].

From Classical Project Manager to Scrum Master

Software development in a fictitious company has been based on the classic waterfall model. In the future, a more flexible approach is required, which supports short-term and incremental deliveries to customers with equally short-term customer feedback. Therefore, the Agile Operational Framework Scrum is to be used as an alternative. In order to cover the increased demand for Scrum Masters, those Project Managers with qualifications in project management, have received additional Scrum Master training.

For some of the Project Managers, this involves little more than learning a range of words and terms, since in classical project management, the focus is exclusively on management processes. In other words, ensuring that each individual in the project is able to fulfill their specific role and responsibilities: The software architect is responsible for architecture, the web designer for web interfaces, the test manager for testing, etc. This means that the Project Managers themselves have minimal intervention in the design of the respective types of results, but assign roles and make decisions. For the majority of Project Managers, however, training as a Scrum Master, responsible for compliance with Scrum rules and the elimination of disturbances (such as conflicts) and obstacles, this means a change from being an instructive leader to a leader serving the team.

In order to exchange ideas informally, Scrum Masters and Project Managers who are new to this role, meet regularly in a nearby restaurant.

Harry Jump, a very experienced Scrum Master, who started his professional career as a software developer in classical project teams, enters the restaurant shortly after the agreed time. He is small and stocky, hair thinning on top, and walks with a confident and relaxed air. When he speaks, his body mimics his attitude.

Inside, Christopher Sedate, an experienced and respected Project Manager is sitting at the bar. He has only recently taken over his first Scrum Team, and in his project environment he worked so far, the Project Manager was an important decision-making body. He is talking to a few other colleagues whilst holding a beer. His stern face and white, combed-back hair, make him look older than he really is. Swing music from the thirties is playing in the background.

Jump notices Christopher, in whose project team he has worked in the past as a software developer and sits down next to him at the counter.

"Well, how are you finding life as a fledgling Scrum Master?" he asks Sedate.

"Well, you know how it is, a new buzzword every couple of years. Yesterday, we were all working according to strict processes and today we are all agile. Who knows what tomorrow brings," Sedate responds wistfully. He continues: "I have already run many software development projects and was initially a developer myself. To date, I have been able to give my team valuable tips for developing. But now as a Scrum Master I am being asked to merely watch the development. Yet I know much more than most of these still inexperienced developers."

Jump smiles to himself remembering his own positive experiences with Sedate. He replies sympathetically, "Well, you know, it's like your children. How can they ever become successful if you do everything for them?" "Yes, you're not wrong," admits Sedate dryly. "But my children do not have delivery due dates looming!" Jump replies in an understanding tone: "That's true! But if your team 'can stand on their own two feet', it will save you a lot of work and you will be able to concentrate on your new role. The team should be able to self-organize. That's the basic idea."

"How is that supposed to work," Sedate counters: "In the past, I've always had all the key information and assigned roles to tasks. If everyone is now going to organize themselves, everyone will need that information and it will require enormous effort!" Jump considers that his colleague has not yet internalized the background information and responds: "Scrum resolves the central role of the classical Project Manager by shifting their responsibilities onto several shoulders. At its core: Responsibility for development results to the team, and responsibility for complying with Scrum development process to the Scrum Master. This also distributes the required information to several people, who will coordinate daily within the Scrum operational framework. All information is available to everybody, including the work steps of other team members. However, each individual only has to gather information to the extent that is necessary for their own work as part of the whole. At the end of each sprint, the Scrum development process is then screened in the so-called Retrospective by the team members, to identify weak points, and changes, if applicable, are decided on together. During a sprint, the current Scrum development process is the basis for each individual's actions, that means, that the simple rules contained within it are sufficient."

"Let's take the example of a flock of birds," Jump replies, arousing his colleague's curiosity. "Do you think that within a flock of birds, each individual bird takes the 'flight data' of all of the others in the flock into consideration for its own flight path?" "Interesting example," Sedate replies, now with his attention piqued: "No, I can't imagine they do!" "Exactly!" exclaims Jump. "Each individual bird requires only a very few parameters for its choice of flight path, such as the flight speed of its neighbors and its angle to them, that is, local perceptible 'data'." "Great! And how does this help me or my team right now?", interrupts Sedate to stop his colleague and attempt to bring him back to the original theme. Then adds mockingly: "So what about my two parakeets at home who just in a cage all day and have no flight formations."

"Well, two would not be enough for a formation," counters Jump jokingly, realizing that he has insufficient background information to explain further. "But, seriously. There are also analogies in humans. The best one I know is road traffic. If we disregard navigation systems, traffic news, and guidance systems, then we can see that in order to get from location A to location B, we only need to apply a handful of rules, 'stop at a stop sign', 'red light means stop', 'give way to vehicles at a roundabout', to name just a couple. None of us know the 'driving data' of the other road users. Primarily, we only need to know the limited driving data of local road users, from our point of view, in order to be able to apply these few rules. As long as certain conditions are met, everything functions by itself."

"Interesting analogy!", Sedate admits and listens more attentively to his colleague. Jump senses this and keeps on going eagerly: "And in a certain way, Scrum supplies this framework, as well as some of the rules. In a metaphorical sense, everyone is able to travel independently from A to B in relation to their current task. The product owner determines the final destination of the 'travel group', i.e. the focus towards which everyone works together, and the Scrum Master ensures compliance with the 'traffic rules', the details of which were, of course, worked out retrospectively by the team. And I would understand the 'traffic rules' as including more than just the rules of the Scrum operational framework. The Scrum Master is not only the coach for the process, but also for communication and collaboration in the team. Important parameters in this context are the workload and appreciative communication between the team members."

"And everyone just complies with these traffic rules?", Sedate asks him. "Why do most people on the road follow the rules?" The ball is in Jump's court. "Hmm. Sometimes, it is actually an advantage to break the rules," Sedate queries the analogy proposed by his colleague. "For instance, at night at a set of traffic lights, why should I wait if there is no one else at the crossroads?" "That's exactly the point," Jump interrupts and explains: "The traffic light controls you, regardless of whether there is traffic or not. Just like you used to control your team, according to fixed specifications. Scrum, on the other hand, behaves more like a roundabout: The latter gives the framework, but you decide for yourself how and when you go, of course, while following the rule that traffic already on the roundabout takes precedence. Traffic lights, on the other hand, unnecessarily lengthen your travel time, yet ignoring them, poses a safety risk!"

"But once again the question arises: Why do the vast majority of people on the road follow the rules?", Jump insists, since his colleague's initial response had not taken the conversation in the direction he wanted. Sedate replies thoughtfully. "Well, as you mentioned, safety. But, of course, we would have chaos in the streets without any rules, no one would ever get to their intended goal," "Exactly!" exclaims Jump and proceeds with the explanation he already had in mind when he first asked this question: "Traffic rules are definitely perceived by the majority of the road users as a tool with which they can achieve their own personal (driving) goals. And that is why most people stick to the rules."

"So, you are telling me that I have to make sure that every single member of the team keeps in line with the team's 'traffic rules', and not only because this is what Scrum demands, but in addition, I also have to ensure that team members are aware of the fact that it is to their advantage!", Sedate summarizes.

"Yes, you could say that," affirms Jump. "Let's take an example of one of your developers with a new idea of how to perform function tests faster and easier. Ideally, they should have the opportunity to explain this idea to the team, and the team could then decide on its implementation together. This increases its acceptance and meaningfulness."

"You are saying that they are more than just a little cogwheel in a big system, as they are in Taylorism," Sedate interrupts. "Can you give me another example of how Scrum operational framework can meet the basic personal needs of team members?"

Jump continues excitedly, "Yes, of course. For example, with Scrum, moving away from a one-time delivery in the waterfall model to incremental smaller deliveries, also means a higher level of probability for a higher number of success stories in the team. The failure rate will even be much lower," declares Jump firmly before rounding his argument off: " … and this strengthens an important psychological basic need: self-appreciation! Especially when the result has been autonomously developed by the team! Which in turn also meets the need for orientation and control. On this basis, rules must be worked out retrospectively." "Yes, as you have already mentioned. It reminds me of my children, who are now at an age where they want to go their own way and make their own success in life," Sedate reflects thoughtfully.

Jump is persistent as he wants his colleague to reach the goal. "You also have to ask yourself honestly whether, when you give instructions, it is actually to increase your own self-efficacy. For someone who can dictate to others, this is an effective lever. The others, in contrast, are deprived of the opportunity to experience this feeling and to satisfy the corresponding basic need." Sedate suggests: "You mean, the bread that you eat yourself, cannot satisfy another's hunger."

"Figuratively speaking, yes," says Jump. "And so you have discovered the central point of the role of the Scrum Master: By not giving omniscient instructions, you transfer much more space and responsibility to the team, giving it 'a niche' for its basic needs."

Sedate adds quickly: "As you have already indicated, the point is to find a task for everyone in the team that is aimed at the project goal and which also provides them with 'something'. In the end, the Scrum Master has to 'adjust' the team so that it acts as a unit for the implementation of customer requirements, that is the trick … I understand!"

"And this has the potential to be even more interesting than software development, right?", Jump concludes, moving excitedly at the thought of having convinced Sedate of the advantages of Scrum, and spilling half a glass of beer in the process.

5.3 Fluid Organization

Organization and environment, or in other words, system and context, form an interactive relationship. In most cases, context, i.e. the market or customers, is likely to be much larger and more important than the organization, so that the organization must adapt to the environment in order to survive.

If the environment is still quite stable, the value-creating tasks of the organization and the associated structures and processes are also unlikely to undergo changes. Transformation work is also rarely demanded, even from those who work in such an organization.

If the environment becomes more volatile, it is necessary for the organization to have or to develop skills that enable it to interact appropriately with the context. Our observation is that in future, this ability of flexibility and speed (therefore agility) will play an increasingly important role in a volatile environment.

Projects or other similarly temporary initiatives, such as feasibility studies or pilots, are possible types of response for dealing with unique questions in (outsourced) temporary organizations. Those in the project team who deal with such questions, still remain legally within their parent organizations (legal entities). In most cases, they also work there for the majority of their working hours and only spend a small portion of their working time on the project. This results in known conflicts of interest between the project and line organization. While money is earned in the line organization with more or less routine tasks, money is spent on the project, and the stakeholders of the project are challenged with new ideas. It is only after a certain period of time that the degree of novelty decreases in the temporary organization "Project".

Figure 5.4 illustrates this relationship: On the left is the parent organization with a hierarchical organizational structure. Dilts Pyramids are clearly shown in different colors. This is to illustrate symbolically that in principle, each sub-organization should be modeled by a different Dilts Pyramid. Compared to the project organizations, the parent organization carries out routine tasks. The project-oriented temporary organizations are shown on the far right in the figure: The team members of the project organization come from the parent organization and, as required, from partner or customer companies. In any case, the project will develop an identity, another Dilts Pyramid, independent of the parent organization.

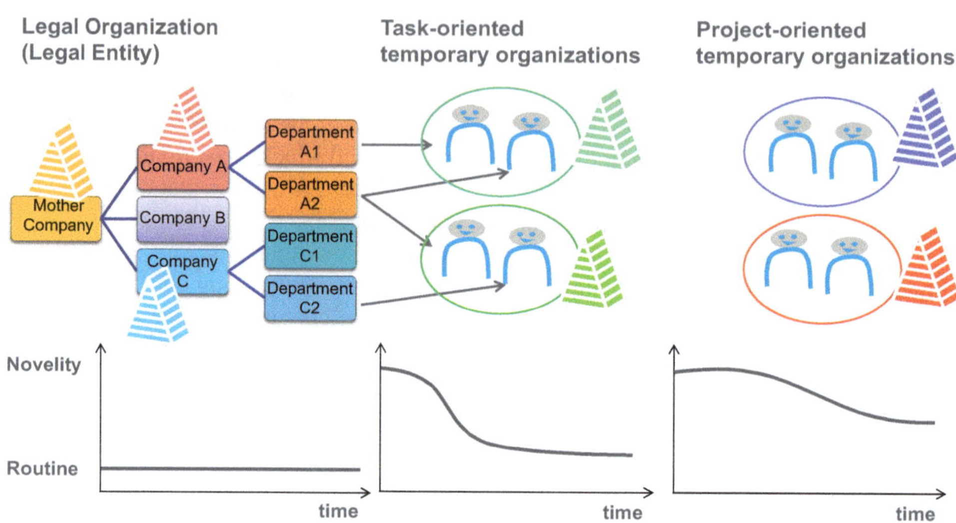

Fig. 5.4 Legal (line) and temporary organizations

In addition to the parent organization and the project, in today's economy a third form of organization, which is temporary, but can last much longer than a project, begins to develop. This temporary organizational type is, for example, suitable for tasks such as establishing a new sales channel or a new product line. Initially, such tasks will still have a high degree of novelty, but after a relatively short period of time, a routine-like state will occur. The employees of this task-oriented temporary organization remain legally in their respective parent organizations and are only lent to the temporary organization. This will be for as long as it makes economic sense to carry out the task in the task-oriented temporary organization: If the sales channel or the product line no longer makes sense, the task-oriented organization is closed and the employees return to their respective parent organizations or develop other products or services in new task-oriented temporary organizations. Depending on the volatility of the environment, this process of setting up and closing task-oriented temporary organizations will be repeated.

We talk of Agile Organizations, if the organization is able to adapt flexibly and quickly to its environment, without setting up and closing task-oriented temporary organizations. We talk of Fluid Organizations, if there is an organizational network consisting of parent organization(s), task-oriented temporary organizations and project-oriented temporary organizations, which dynamically adapt to the environment.

Therefore, a Fluid Organization must significantly increase its ability to transform, otherwise friction between the different organizations will lead to the opposite effect. Instead of flexibility and speed, value-destroying complexity develops in all forms of expression: emotional blockages, stress, extra work, …

The models necessary for the development of the ability to transform, were introduced in the chapter "Leadership in Complex Social Systems". Of course, the design and application of organizational transformation is always organization-specific, so at this point we will point out some consequences of a general nature:

Those individuals who have had no opportunity to expand their ability to transformation in task and project-oriented temporary organizations, will hinder transformations. This applies in particular to executives who "always" remain in the parent organization.

Fluidity can only occur if the manifestations of the upper levels of the Dilts Pyramids of the various organizations are not too far apart. For this purpose, in an ideal case scenario, it is necessary to pass, through the transformation process described in the section "Transformation Management", for each transformation and with all those involved. Let us start with the transformation step "illuminating pressure points", which corresponds in this case to illuminating what is needed for the structure of the task-oriented temporary organization. The setting, control and order parameters of the new organization are then worked out. This involves checking whether in the future, the new system parameters can lead to problems in and with other organizations. This measure corresponds to the "Future Pace" known from NLP, which is applied to organizations. The system parameters are then subsequently detailed for fields of action. Care must be taken to ensure that the system parameters, transferred into guidelines, are moderately detailed, and always following the basic assumption that behavior should not be prescribed, as this inhibits the development of agility and fluidity. The final step in the transformation cycle is the deliberate introduction of structures that support higher-level learning (learning stages III, IV).

Fluidity Begins in the Mind

Fitzherbert, a tall, mysterious looking man wearing a hat, which he wears in an odd fashion, pulled slightly down over his face, hopes that his appointment with the IT service provider is just routine.

"We have the opportunity to win a group of new end customers in the market," says Fitzherbert, looking specifically opposite at the slimly built key account manager Dennis Lee, on whose IT services Fitzherbert's company is so dependent. "Lee listen, in the short term, we need a new database-based application software," he continues calmly, emphasizing each word, "which should be serviced through its entire life cycle, by one single provider. Is this something you can do?" he finishes.

"Yes … certainly," replies Lee, initially with slight hesitation, but then continuing in a more assured tone, "We work mostly in task and project-oriented temporary organizational units. Although the employees formally belong to departments within classical line organization, these task-oriented organizational units often continue during the entire life cycle of a product. On the other hand, project-oriented organizations, i.e. projects, only exist during specific product phases, and for which they receive an order from the task-oriented organization for their implementation. Task-oriented organizations also consist of the roles responsible for the entire product life cycle, e.g. the customer-, product and release manager. You could say that line, task and project-oriented organizations form, in this order, an arch from long to short-term stability: From basic company policies in the line to long-term orders in task-oriented organizational units to their further development in projects." Lee wraps up his uninterrupted monologue.

"Well, that looks great on paper, and you've explained it quite well," Fitzherbert nods, not without noticeable appreciation. He then adds: "Another indication that line and project management merge slowly."

He continues: "How do you ensure that the temporary cooperating specialists from the different line organizations function as a cohesive team?"

Lee replied quickly: "I suggest you participate in the kick-off meeting for the build-up of the task-oriented organization. In this meeting, database and user interface experts from two specialized line sections will also be present, as the initiation of the first product phase, i.e. the development project, will also be discussed."

Fitzherbert is once again impressed by the seemingly well-informed Lee and places a temporary limited order. He even accepts the invitation to the kick-off meeting.

In addition to Dennis Lee, the product manager, the release manager, developers of the databases and user interfaces, are also present at this meeting, in addition to Fitzherbert and his team of experts on the business function of the newly developed application. After Lee briefly explains the background of the structure of the temporary organizational units, he hands over to the manager responsible for the new task-oriented organizational unit, Mark Hughes.

Hughes, a small man with a look of Inspector Columbo, introduces himself and then addresses the auditorium: "Could you please briefly describe those factors, which in your opinion, are the most important for a successful new development? And, furthermore, I would like you to tell us why you believe these are important." Frank Davies, a user interface developer, a man with a goatee beard and horn-rimmed glasses, begins: "As a user interface consists of dialogue with input and output screen masks, we need to create a prototype at the earliest possible stage in the development process. This is the only way to get feedback from the end-user at an early stage and make appropriate adjustments." "What happens if, at this early stage, there is still an error in the business functionalities of the application?", Hughes asks. "Of course, we are working in a test environment during the first stage, but even errors in the production process can lead to limited financial damage if, as recently happened, an incorrect order was triggered with ERP software," confirms Frank Davies.

Derek John, a database expert interrupts: "We are mainly developing databases for applications in operating theaters in hospitals. Errors in these applications can risk life and limb. This is why every single intermediate step, e.g. software module, is tested separately during the development process and has to be approved before carrying out the subsequent step. A test with all submodules of the application only takes place at a later time therefore."

Fitzherbert looks approvingly and repeats: "Risk life and limb… what other kind of software do you work with … ?"

"Good to know how diverse experiences can be," Hughes retorts quickly hoping to move on and directs the conversational focus to one of Fitzherbert's experts:

"What are the requirements for our new application?" "A mistake in the new application can lead to high financial losses, but are no threat to life and limb," chuckles Dorothy Soberg, one of Fitzherbert's experts for the business functions of the new application to be developed, and then continues with a more serious expression: "However, in the case of an accumulation of errors, the business case would no longer be valid."

"That should be enough for us," concludes Hughes, turning to the developers: "What do the user interface experts say about the highly detailed and elaborate development processes described by the database experts? Will the processes result in a significant reduction in errors in applications?"

"Certainly. However, I consider the process to be completely over-the-top. If we had to work with them, they may overshadow our creativity. Or, our creativity will be used to build more and more formal release processes … and I do not believe that is what we want," argues Frank Davies.

Hughes feared such an answer and Fitzherbert stared ahead expressionlessly murmuring to himself: "Ok, so we are not really getting anywhere."

A conversation ensues between Hughes, Lee and Fitzherbert: "I was afraid of that," Fitzherbert begins, "There are so many different schools of thought that are clashing. One of them views IT as a colorful children's playground and wants to be creative, yet for others IT is more of a martyrdom self-imposed of hurdles, where everything has to be carried out in the most precise manner."

"Well Mr. Fitzherbert, there is certainly something to your analysis, but you also have to admit that different requirements lead to different ways of working and subsequently to different value propositions. If different departments clash in terms of key questions, and mine was one such question, then controversial discussions are quite possible. And this brings us to the core of the matter: Our corporate culture enables us to speak openly about dissent. This is the best way to recognize and constructively deal with it. Constructive in the sense of resolving any dissent concerning the facts or the relational level, and preventing different perspectives leading to personal offense and subsequently to conflict. It is therefore necessary to take into account both hard and soft factors," retorts Hughes.

"And all this in the presence of customers …", grunts Fitzherbert.

"This is exactly what demonstrates our willingness to transform our company: Instead of sweeping dissent under the rug, it is addressed in advance. And our employees also have the courage to do this in front of customers. This allows the customer to intervene in the design of the co-operation at an early stage, provided they have the necessary know-how," Lee interrupts.

Fitzherbert turns to Lee: "But don't you think I would not already be aware of this Mr. Lee. I would initially ask whether the two developers, Mr. Davies and Mr. John get along with each other on a personal level, without being influenced by their departments."

Following this conversation, other topics are discussed, before Mark Hughes leaves the kick-off meeting with 'homework' muttering to himself: "We have a task-oriented organization, yet have to initiate the necessary social techniques for transformation."

Initially, Mark Hughes lets Derek John, the developer with rigorous test procedures, collide with Frank Davies, who is more creative. They both get along with each other from the very start and even begin to develop a web interface, which is to be tested with elaborate procedures. Davies and John, are able to see whether they are compatible with each other. However, the good atmosphere does not last long, as Frank Davies' creative colleagues get wind of this development and fear that from now they will also be obliged to test super accurately, and maybe even experience a hostile take-over by the database developers. Hughes is unable to dispel the concerns of the creative colleagues despite various discussions.

Fitzherbert is on the verge of dropping the project, when during an evening walk, he has a flash of inspiration: "We want a task-oriented organization with a coherent mindset: so we can eradicate group thinking by assigning Frank Davies to the database developers. I must talk to Hughes about this tomorrow."

Fitzherbert and Hughes meet first thing in the morning. Later on, Hughes chats to Derek John and asks him: "How do you rate the perception and acceptance of database developers in our company?" "Well, we are not really appreciated. Our software clearly works better than any other … if you want to be effective, you need to be creative", Derek John thinks to himself.

"Can you stand a little more creativity?", Hughes adds: "And most importantly: how will your colleagues take it?" John laughs: "I think I understand you. Yes, then let's take in Davies, he may well mix up our database business. This might be good for the team. And he is certainly not lacking enthusiasm for sharing his creativity. I'll make sure he doesn't get carried away! We actually get along with each other quite well", Hughes thanks him heartily and informs Fitzherbert right away.

Now it's Davies' turn to drive forward database development with his creative mindset. He also has experienced, quality-conscious colleagues breathing down his neck, who are curious as to how to develop creative surfaces, and subsequently perhaps be appreciated a little more, even as high-level, quality-oriented database developers …. the best of both worlds ….

In one of the following requirement workshops, in which Derek John, Frank Davies, database experts and interface developers are participating, again, despite Hughes' previous interventions, there are again two different camps questioning how detailed the development process must be and how many tests and releases are required. As a consequence, Hughes assists in the workshops and provides support. He recommends using the three-tier model of the Collective Mind Method.

He therefore introduces the next workshop: "I would be interested to hear what you think of my proposal, the Goal-tier, which so far only consists of expected customer benefit, to be complemented by further levels of detail and we can do this by paying attention to the product life cycle, as well as to the objectives for the development project? At this point, we should also grapple intensively with our beliefs and values."

Hughes continues: "As bugs in the application result in financial damage, I suggest that it is merely a question of how our customer's profits, that is the company Mr. Fitzherbert represents, can be maximized over the product life cycle. In other words, the cost of a complex development process must not be higher than the operating phase costs that would have been caused by software errors as a result of a less complex development process. And regardless of whether the costs are caused by the process or by errors, they must be taken into account in the business case. In order to provide an information basis for this decision, I propose creating a common picture by developing a tier for functional architecture as a next level of detail, followed by a further tier with the software modules that represent these functions. This should be done within the next two weeks, because the entire development project should be completed in about six months."

After about 2 weeks, the goal picture ("Goal-tier"), the functional architecture tier ("What-tier") and the software modules tier ("How-tier") are presented. At the "What-tier", the business functions are divided into "business-critical" and "standard". "Business critical" means high or frequently occurring financial losses in the event of an error. "Standard" means moderate or rarely occurring financial losses. By making explicit the belief systems, the following convergence emerges: Initial accuracy in the OP-area is no longer necessary in the context of the new database application. This results in agreement amongst all experts: The software modules ("How-tier") are classified as "business-critical" or "standard" depending on the respective assignment of the higher-tier business functions ("What-tier"). The number of tests and releases of the respective development process depends on this. A project manager, still to be determined, will have to take into account the resulting different degrees of management: The activities for a "business critical" class module should be much more finely planned and traced than in the "standard" case.

With the help of a robust, jointly developed hierarchical target picture of the initiation and strengthening of appreciative communication, and the integration of Frank Davies into the database developer group, it was possible to break down barriers much faster.

Consequently, and taking into account beliefs and values, the application of the Collective Mind Scheme, has resulted in a technically simple solution (minimizing technical complexity), while at the same time developing an expert group into a team (minimizing social complexity).

On completion, the experts, in particular Davies and John, are curious as to whether Hughes studied his approach, or whether it was simply down to professional experience.

"It can all be learnt," replies Hughes, knowing that a combination of professional experience and social techniques was what had helped him acquire these skills over time.

During this conversation, Hughes was even able to convince the team to take part in the next training session on the use of the Dilts Pyramid. This was a further step towards the acceptance of social technologies, as he practices them. He feels justified in his actions, and the idea of the "fluidity" of the company has once again developed a little in their minds. Construction of the next task or project-oriented temporary organization will not only be easier for all parties, but may be welcomed by his colleagues as a pleasant change in their professional life. And customers will also appreciate it.

5.4 Virtual Team

We talk about virtual teams, if the majority of work in the team is carried out using electronic media such as telephone, e-mail, video conferencing or social media platforms, so that the team never or rarely meets face-to-face. Virtual teams can occur in all organizational forms.

Figure 5.5 shows an example of a virtual team consisting of nine people who feel at home in different organizations. We therefore also talk of five groups, which can represent up to several hundred stakeholders, depending on the size of the organization. For the sake of simplicity, all the organizations belong to one national division within a company.

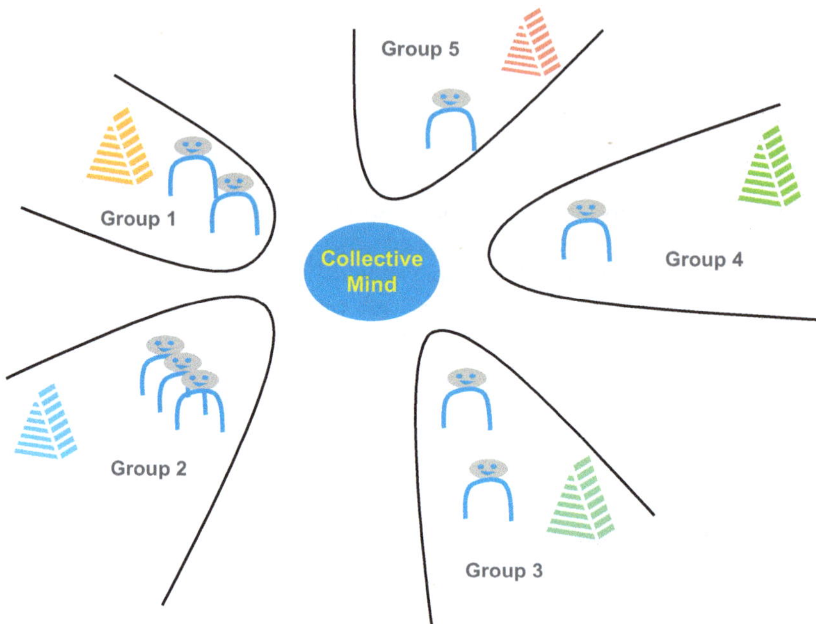

Fig. 5.5 Virtual teams

In our example, we have assumed that the team members have never worked together on a project. The project manager belongs to group 5 and as part of the project initiation, has classified the project as a missionary project from the customer's perspective (high degree of novelty for the stakeholders and high degree of innovation). Because time is money and all team members are very busy with their daily business, the project is to be carried out virtually from the very beginning.

And how is the project coming along? Most readers will probably conclude that the project is likely to end in disaster. But why is this, what are the decisive factors, and how can they be successfully shaped? Or in other words, how much virtuality can a project tolerate, based on its project setting, the environment and the project dynamics?

In our example, we have assumed that two of the organizations (group 1 and group 2) can be characterized by the interaction example in section "Interaction Between Micro- and Macro-tier" (see also Fig. 4.20). There, we had tacitly assumed that teamwork was carried out mainly in face-to-face meetings, and we were able to show that non-coached interaction was likely to lead to considerable problems in cooperation. Let us take a closer look at interaction in a virtual team: Group 1, for example, is described as an IT department with a polychrome and developing understanding of time, and group 2 represents the business department with a monochrome and planned understanding of time. In the first virtual meeting, no problems related to these differences arise: All turn up on time for the video conference. But for the next meeting, Fred Landau requests that he only take part by phone, as his presence is required for a very important system test. Ms. Sally Harris, as representative of the business department (monochrome and planned understanding of time), begins to get angry, but does not say anything yet. At the next virtual meeting, this process is repeated. Ms. Harris insists on a binding commitment for the completion of an IT task, but receives only an evasive answer from Mr. Landau, as the subject was complex and he therefore could not foresee a possible solution.

This example demonstrates that virtuality can, in some ways, significantly increase existing problems. In our experience, this is most likely to be true of all problems; in some cases, problems that were latent suddenly become visible. We recommend that the reader to be aware of the effects of this setting on a virtual team by using the example above.

We assume therefore that virtuality facilitates the creation of value-destroying complexity. The creation of value-creating complexity in the form of self-organization is not promoted in our experience. Working together in front of a flipchart or whiteboard to create a Collective Mind can, in our experience, not be reproduced in a virtual team. Unity of place and time, the related exclusion of organizational complexity of parent organizations, and the creation of a sense of community, are achievable only to a limited extent. This sense of community is a colloquial expression for the creation of an organization-specific mindset (Dilts Pyramid) of a temporary organization.

We therefore recommend transition to a virtual team, only when a sufficiently stable Collective Mind already exists, i.e. the Tuckman phases of team-building have been completed and possible interaction blocks have been tackled through transformation work. As a rough timeline guide, we recommend a 2-week period of joint and intensive team work at the start of the project. If necessary, the Collective Mind should be refreshed by a face-to-face meeting after 3–4 months at the latest.

Virtuality Reinforces the Impact of Local Influences

Communication between Fred Landau and Sally Harris has now literally collapsed, and cooperation is no longer possible. This is due to the different departmental cultures in the IT (i.e. developing, polychrome time) and the business department (i.e. planned, monochrome time) and the different individual temperaments of the IT employee Fred Landau (Champion: ENFP) and the business employee Sally Harris (Supervisor: ESTJ): Conflict manifests itself first and foremost in communication between the two colleagues. Working in a virtual team has greatly intensified the conflict. The only way out is by bringing in a coach trained in social techniques, who can make the effective system forces transparent and stimulate transformation work.

Don Schmidt is such a coach. At 1.8 m and the temperament of a bad-tempered John Wayne, he greets Fred Landau and Sally Harris with a nod. He is someone who enjoys provocation and by this means intentionally creates extremely effective transformations. He starts quite directly: "We will turn this shop upside down to provide you with the necessary insight into effective forces. You will need to take a great leap into the unknown. Do you think you are ready for this? There's still time to turn back!" The corners of his mouth curl as he speaks.

Sally Harris and Fred Landau, although visibly irritated, agree. "Good. I will start with individual interviews," Schmidt finished the conversation. All three say goodbye. Ms. Harris and Mr. Landau walk together for a while. "Odd character, this Schmidt. And strange manners too." Fred Landau says, shaking his head. Ms. Harris replies "But in some ways refreshing. We can still bring him in line. Are you on board?" They say a friendly goodbye, somewhat relieved.

The next morning, Don Schmidt and Fred Landau meet for their individual interview. Fred Landau, at first seems lost in his thoughts looking at his appointment calendar. "Are you a mere paper-pusher, or what?" Schmidt snarls at him. "No … no, I'm just looking at my calendar," he replied, slightly offended. "So, this vital meeting is not of the least importance to you!", Schmidt asks unceremoniously. Landau is speechless and is about to abort the session, when Schmidt quickly stops him with a wave of his hand: "I see you are one of those gentle, creative people. With your consent, I would like to work with you on a questionnaire about your temperament, to discuss typical, preferred behavioral patterns with you. However, we must also discuss your department," adds Schmidt. Landau suddenly becomes more interested and starts to introduce typical IT department workflows. He is enthusiastic: At the end of the meeting he knows his temperament somewhat better, as well as the basic values and principles of his own department. He also understands that his temperament fits the culture of the department. In reply to Schmidt's final question, "How do you want to work with Ms. Harris in the future," he replies, "Of course I would like to tinker about working on computer programs. But, I am also aware that Ms. Harris needs solutions." "And what do you expect from me?" follows up Schmidt. Landau responds sensibly, "A restraint system that reminds me to stay on track when my instinct beckons me to play around." "That, we can do," wraps up Schmidt. With a short bow he says goodbye to Landau.

The individual talk between Mr. Schmidt and Ms. Harris is completely different. She meets him with a firm, challenging look and Schmidt notes that he will not get far with his provocation technique. In a very factual, but friendly conversation Ms. Harris, in a similar way to Mr. Landau during his appointment, gets to know herself and her department. Naturally, she considers her own temperament and her department culture to be the most correct, and she expects precise legwork from Landau.

In addition, both Mr. Landau and Ms. Harris note that Don Schmidt's intention is not to change their personalities to resolve conflict, but to understand their personalities along with the values and principles of their departments, in order to emphasize their strengths and compensate for their weaknesses.

In the first meeting between Fred Landau, Sally Harris and Don Schmidt, the latter begins in his usual provocative manner: "It seems that recently you have worked together really well!"

They both remain silent for a moment, until Sally Harris retorts: "Mr. Landau always seems to have more important things to do than work on our joint project." "Well we have a lot of work to carry out simultaneously," replies Mr. Landau apologetically.

"Thank you for the excellent feedback," Schmidt remarks dryly: "I will interpret your comments as a clear no. And that is the reason why we are sitting here together: I want us to work out together, how your different worlds 'tick', including your personal temperament and departmental culture This will provide you with a model, which will be a key to the solution. Would you like to lay your cards on the table?" Both are silent for a moment, then agree with a nod. "Ms. Harris, do you agree?" encourages Don Schmidt. "Well I think we are already aware that we have different views," replies Sally Harris. "And Mr. Landau, you nodded, but how would you express this verbally?" "I agree with Ms. Harris. We have different expectations from each other," replies Mr. Landau in a tone of relief, as if he had wanted to say this for a long time. "So, I think you both agree that you have different views and different mutual expectations. But your common goal is to work together efficiently in the future," Schmidt summarizes their words as the first Goal-tier (Big Picture). Both Ms. Harris and Mr. Landau reaffirm with a sincere 'yes'.

"Mr. Landau, I'd like you to imagine that you are Ms. Harris. How do you think Ms. Harris would describe your views and behavior?", asks Schmidt, in an attempt to build mutual understanding through meta-perception. Mr. Landau thinks out loud: "After yesterday's discussion with you, I am better aware of how I work: I think Ms. Harris would probably describe me as impulsive, unstructured and unreliable. I think that she does not see my creativity." He thinks for a moment and then adds, "... if I worked in a department as structured as the business department, then it would not make sense to approach things in different ways. Everything would need to be processed step by step. And that does not suit my temperament either. I think I now understand why someone like Ms. Harris is much better off in the business department than me and why she perceives my personality in such a way." Ms. Harris is also asked to answer the same question from Mr. Landau's point of view: "Mr. Landau would probably describe me as persistent and extremely structured." She also adds thoughtfully: "A department with polychrome time would not work for me. I need clear rules and decisions. I pity Mr. Landau, who has to deal with this in his department ... but then again, he is also quite different from me, and it may suit him. I also now understand why the IT department works on so many projects at the same time: It has to deal with very different customers. They are quite a colorful bunch."

Both of their answers are then mapped to the temperament and culture model. There is a good match between the statements and the models: What has so far just been 'felt' can now also be rationally described and therefore tangible. A constructive approach to the different behaviors of Ms. Harris and Mr. Landau is now possible, as they have been made explicit and somewhat internalized through role reversal.

"You've talked quite a lot now. Do you think it is only words that matter in communication?", Schmidt queries.

"Body language is just as important ... ," adds Landau, "Good point! What is overwhelmingly important is body language and intonation," Schmidt insists, continuing: "If, for example, what is said, contradicts facial expressions, it is facial expression which often determine how we react emotionally. This is especially interesting, when we consider that we are all so rationally educated. Descartes sends his regards! This makes it easy to grasp the importance of the audio-visual quality of the communication media used for virtual teams. But even the most perfect audio-visual transmission cannot completely replace face-to-face communication. Not everyone, however, is aware of the consequences."

"Yes, you also need to have experienced meeting each other face to face, without which even the most expensive video conferencing will be useless," Landau states and Sally Harris agrees.

"Ms. Harris and Mr. Landau, you have worked quite intensively today!", Schmidt says. He finishes the coaching session and starts a feedback session, before he suddenly stops speaking. Mr. Landau affirms: "Playing the role of the other person in an established behavioral interaction pattern, to be able to see myself from the perspective of the other person helped me greatly in understanding both the Ms. Harris' reactions and my own behavior." He now feels confident that he has found a construct with which he believes he can change collaboration in the virtual team to a better one.

"I now understand and accept a little better Mr. Landau's behavior, even if it needs some getting used to," adds Ms. Harris.

"I also believe that you are now ready and able to return to your environment," confirms Don Schmidt, closing the coaching session and saying his good-byes to Sally Harris and Fred Landau with a firm handshake.

Both Ms. Harris and Mr. Landau now understand that the effect of temperament and local environment is much more important than they initially assumed.

Mr. Landau would like to set a good example and speaks to a colleague about Ms. Harris' future IT task: "Hello Thomas, how far have we got with the new mail service for the business department?" "Fred, you must realize that it's a very complex subject," retorts Thomas Green, "and a possible solution is not yet in sight. Also, this new mail service is only one of several tasks I am trying to handle simultaneously." "I totally agree with you, Thomas," counters Fred Landau, trying to slip into the role of a virtual team representative: "We have a very innovative task developing this project for the business department. So many of our colleagues are already waiting for the new mail service, even colleagues in our own department. And I am also proud of the fact that we have been asked to develop software for the business department, as it is quite important." Landau emphasizes his words as he assumes that Thomas Green would never himself accept the importance of the business department, as he is a true techie. And what techie likes customers…?

"Yes, technically the whole thing is very interesting …," reflects Thomas Green in a low voice. "You know, we have actually agreed a delivery date with the business department and our colleagues there have adapted their plans accordingly," continues Fred Landau: "Did you know that the business department prefers fixed schedules and also avoids working on tasks simultaneously." "All right, I give in." shrugs Green. "It seems you are under the thumb of the business department and have to tolerate their habits, too. I understand. Let's quickly find an appointment for this week where we can further define the task", he concedes.

A week later, when Mr. Landau was already late for the next virtual team meeting and in a rush, he was stopped in the corridor by Ms. Taylor, a colleague from the IT department: "Mr. Landau, I'm so happy to have bumped into you! We found another production-preventing error last week during the system test," she explained nervously. "Oh, I thought we'd eliminated all serious mistakes," Mr. Landau replied, before suddenly remembering his agreement with Ms. Harris to attend the virtual team meeting on time, and quickly adds: "Please can you send me the test report and an invite to an internal meeting with the test and release manager for today. I need to rush off now to a meeting for another project with Ms. Harris."

Mr. Landau reaches the videoconference room just in time. The other participants and Ms. Harris, are already connected. Ms. Harris is a hundred kilometers away in her usual location.

The project leader from the business department joins in from a third location, opens the meeting and immediately asks about the status of the new mail service. The question is addressed to Ms. Harris, as she has formulated the business department requirements and is responsible for the project: "Ms. Harris, what can I write about the new mail service in my status report? Are we on track?" Due to a poor reception and delays in sound and image transmission, group 1, the IT department, feel as if the project leader has a negative undertone in his voice. Group 2, the business department, however, have clear sound and image transmission. Mr. Landau (IT department) immediately reacts with emotively: "You do not have to be so negative. You could show a little more appreciation for the extremely innovative work carried out by both the IT and business department!" Ms. Harris is unable to comprehend of Mr. Landau's emotional outburst, but remembers the Coach's words on the importance of body language and tone in communication and asks: "Mr. Landau, for us, over here, the transmission of both picture and sound is outstanding today. Are you also receiving us all clearly?" Mr. Landau also remembers the Coach's words, recognizes he may have misjudged the situation, and begins to calm down. "We have some disturbances in picture and sound here," replies Mr. Landau. Although, on the one hand, the project leader is slightly irritated by Mr. Landau's unjustified reaction, on the other hand, he is positively surprised that Mr. Landau defended not only the IT department, but also the business department. Whereas in the past, they were more likely to have always rubbed each other up the wrong way. Mr. Landau is now aware that he is more likely to make decisions based on values and feelings, in contrast to Ms. Harris, who makes decisions based on analysis and objectivity. Mr. Landau, is also aware of the need to control his feelings somewhat to be able to concentrate on the project, and ensure that the mood in the project remains productively aligned. "Technology is playing tricks with us again!", Landau interjects in a humorous tone and a friendly smile, returning the mood to positive. He continues: "I suggest we reset the connection. We might get a better connection. I'll do it now." Mr. Landau is also aware that it is not only the technology that needs resetting, but communication, too. He adds, "We'll be back in a minute." The subsequent session is then surprisingly productive. The IT department is able to present a prototype developed by Thomas Green, according to Ms. Harris' basic specifications. She is subsequently impressed at the quick result and traces it back to the fact that Fred Landau has managed to include a bit of business department culture to the IT department. After the meeting, Mr. Landau briefly considers what would have happened if he had failed to pay Ms. Harris any attention or mindfulness: It would probably have ended in never-ending, unproductive dialogue again with 'Yes, but ... No, but...'.

Literature

1. Sutherland J, Schwaber K (2016) The Scrum Guide. The Definition Guide to Scrum: The Rules of the Game. Download at http://www.scrumguides.org/
2. Sayer NJ, Williams B (2012) Lean for dummies. Wiley, Kindle Version
3. Schruff E-M (2015) Lean project management. Bachelor thesis, Hochschule für Wirtschaft und Recht, Berlin
4. Techt U (2015) Projects that Flow: Projekte in kürzerer Zeit. ibidem, Kindle Version
5. Köhler J, Oswald A (2009) Die Collective Mind Methode. Springer, Heidelberg
6. Maximini D (2015) The Scrum Culture: Introducing Agile Methods in Organizations (Management for Professionals). Springer, Heidelberg
7. Peters T, Ghadiri A (2013) Neuroleadership – Grundlagen, Konzepte, Beispiele: Erkenntnisse der Neurowissenschaften für die Mitarbeiterführung. Springer Gabler, Wiesbaden
8. Cameron K (2016) Organizational culture assessment instrument. http://www.ocai-online.com/. Accessed 14 Feb 2016
9. Schneider WE (1999) The reengineering alternative. A plan for making your current culture work. McGraw Hill Book

Check for updates

6 Conclusion and Outlook

Complexity is a universal phenomenon and contributes significantly to the creation of new things in our world. Complex systems can no longer be predictably determined, but sometimes produce something entirely new, which may result in the next stage of development: "The whole is more than the sum of its parts". Complexity can also be destructive - and consequently, "the whole is then less than the sum of its parts".

Complexity occurs when there is a sufficiently high degree of interconnectedness and interaction between subsystems. This is also the case with a large number of today's systems, i.e. networked communication processes, public transport systems or networks of companies, including projects - the trend is growing.

In today's project world, we are at a crossroads. Do we want to continue tackling this increasingly complex project world with standard project management methods, i.e. primarily project planning and reporting, and to consider human interaction as a footnote? Do we want to accept the resulting consequence that projects are "just the way they are" and that anyone who becomes a project manager has pulled the short straw. Or do we finally have the courage to cope with complex daily project work by using models and theories? The nth project planning and reporting tool will still not be able to eliminate or master complexity.

This book provides, for the first time, a framework of social techniques for complex systems. We have aimed to present a symbiosis of theory and practice in which a theoretical approach to looking at things is represented by practical examples, as: "There is nothing more practical than a good theory".

In addition, this book provides the necessary basis from which to learn: Every project manager should rejoice at the opportunity to lead a complex project. No one involved in a project should be afraid when a project is said to be too complex to be manageable. However, this also requires that a project leader in particular, in addition to the team members, are prepared to acquire a completely different form of competence: The project

A. Oswald et al., *Project Management at the Edge of Chaos*,
https://doi.org/10.1007/978-3-662-48261-2_6

manager, above all, must be prepared to work with models and theories, test them in practice, and have the necessary openness to implement both theory and practice, continually adjusting both. Linked to this is the need for a high level of self-reflection, in order to perceive the systemic effects of action through a meta-position and to adapt the intervention accordingly.

What does this mean for you? If you have read through this book, you will develop a sense for complex phenomena. You will start to assess project situations in a different way, using models and theories in order to master them.

Convince those around you. Discover complex behavior together. Demonstrate that complexity is right there in front of you: Simple systems can also be complex, whether technical or social systems: The double pendulum in physics, the paths of which can assume chaotic states, as well as the three-body system of the sun, earth and moon. A leopard's patterns have also evolved through self-organization. Just think about the high number of communication possibilities within a team of three people who are working together on a solution. Despite the high level of professional qualifications, experience ranges from mental flow to complete failure. Or, what about the school teacher commenting: "You will amount to nothing" and the short or long-term effects of this? Alternatively, the opposite is true, if a competent university lecturer in mathematics is incredibly motivating with: "You can do it. I know you can". Although the exercises might be extremely difficult and no solution in sight at first, the result will be positive. Just a small sentence can have a massive impact on future behavior.

The content of this book can also be applied to other areas of life, such as politics. Details are unimportant to start with; as they will result in "being unable to see the wood for the trees", instead of a simple description using a few parameters. Now if you hear: "this particular politician has been successful over the past thirty years, so they will be able successful at solving the current problems we have!", then the appropriate system parameters for the following questions should come to your mind: What is the context in which the politician lived? What values do they have? What motives? What visions? Moreover, what were the social conditions at the time? In addition, can all this be simply transferred to the situation today? You will then be able to assess whether there is the probability for a stable macro-state, i.e. if a successful solution can or cannot be achieved with this politician.

If we have encouraged you, the reader, to take on complexity, to look at the world with this perspective, and to potentially simplify your daily project work, then this book has served its purpose. You will begin to see that through continual application of our framework of social technologies, intuition for complex problems will steadily develop and you will subsequently be able to master new challenges and projects.

Appendices 7

7.1 Fundamentals Theory and Practice

In the section "From Negative Words and Basic Assumptions", we postulated that theories and single models must have practical benefit, and thus have to prove themselves in practical application. In many cases, theories are denied this relation to practice and so remain gray. If, however, theories can be combined with practice, both theory and practice begin to shine. Everyone is familiar with detective novels, where the inspector and his/her team, develop, based on first facts, a variety of possible theories, and then little by little, eliminate all the wrong theories after factual verification, until finally, a correct theory remains.

Without this mechanism of collection of facts, combined with the formation of theories, there would be no success in crime investigation, and in addition, it would also not be much fun for the reader. The fun for the reader is in being able to theorize and independently reconcile theory and practice.

We believe that practice without theory is the basis of unconscious incompetence, and theory without practice is a meaningless activity. The idea for this book originates from the joy and curiosity of combining the theory and practice of social techniques in order to end up with new insights.

In the following section, we will explain in more detail, our understanding of theory and practice and use Einstein's theory of relativity as an illustration. We have chosen this theory, because according to common opinion, it is a prime example of a theory, which has far-reaching implications for our lives.

Our explanation of "theory and practice" is a model of models, i.e. a meta-model. This may be difficult to digest for some readers, but at no point do we presuppose anything beyond basic school knowledge.

Our meta-model "theory and practice" is based on the assumption that "nothing is as practical as a good theory". This guiding principle summarizes the insight that we as

A. Oswald et al., *Project Management at the Edge of Chaos*,
https://doi.org/10.1007/978-3-662-48261-2_7

humans must first have an idea of something, in order to recognize something in our life. Perception and attention are very much shaped by our awareness of what we shall, want to or are able to perceive.

It also expresses the fact that a theory, which fails to be substantiated with practical proof, may be interesting, i.e. having an interesting construction of ideas, yet lacking in essentials. In [1] we have used Merton's classification for a theory's range: grand theory, middle range theory, micro theory.

The range of a theory measures whether, and to what extent, a theory is operationalizable and how large the domain is that the theory claims to explain or claims to understand.

Operationalizability means that tools exist, which support the transition to practice. If a "good theory of practice" exists, the models and methods of the theory form the basis for the tools and support the checking of the theory in practice. The theories which we are discussing here are middle range theories. Such theories may be far ahead of their time, and therefore can only be checked in practice at a later point in time.

Observations mediate between theory and practice. Observations can be qualitative descriptions of an experienced practice (stories, case descriptions, or similar) or quantitative descriptions that are carried out under very specific conditions and return verifiable quantitative results. These include experiments, studies conducted under controlled conditions, or so-called field studies in which, according to strict scientific methods, data in practice are collected.

Observations contain one or more phenomena. Phenomena are events that are attributed to an observation and which particularly captivate our attention in the context of the observation. Throughout this book, we have pointed out that attentive observation and associated mindfulness are prerequisites for theory formation and, on the other hand, that theories enable careful observation. In this case, theory and practice are in evolutionary developing symbiosis.

A middle range theory contains statements about practice, so-called hypotheses, which are checkable in practice. If such hypotheses are not present, the theory remains pure theory and we believe therefore, not a good theory. For example, if we ignore the takeoff and landing of an aircraft and assume that the flight is otherwise uniform over a sufficient time period, a difference between flight time measured in the aircraft and flight time measured on the ground should be ascertainable. This hypothesis was confirmed in the 1971 Hafele-Keating experiment and is based on the Theory of Special Relativity.

Linked closely to the hypotheses is a further central requirement to a good theory, namely its ability to go beyond previous practice and deliver prognoses. At the beginning of the twentieth century, based on experimental observation, it was found that the speed of light in the direction of the movement of the earth and away from the movement of the earth was the same.

Einstein concluded that the speed of light is a constant which does not change in linear, uniformly moving systems (so-called inertial systems). He postulated his famous Theory of Special Relativity and predicted on this basis that time passes differently in differently moving inertial systems. In this case, theory followed practice, or in other words, it was formulated after the experiment had been carried out.

In the Theory of General Relativity, on the other hand, Einstein initially came up with the idea of a model from which he derived hypotheses. Later, these were proven in experiments. Following the hypotheses of the theory of General Relativity, things were then found and proven. These include, for example, the deflection of light by large masses such as planets or stars, or the different ways that clocks move in an inhomogeneous gravitational field: Without considering this effect and any associated corrections, no navigation system would work. The gravity of the earth is lower at satellite height than on the earth's surface. This means that the way clocks move at satellite height is faster than on the earth's surface, which needs to be taken into account for any navigation system.

Until now, we have used the pair of terms "theory and practice". By the term "practice" we associate actions that have been "proven" in our private and professional life to have a desired effect. The term practice thus expresses the fact that the actions carried out have shown themselves to be fit for our survival, or the survival of an organization.

In science, the terms "reality" or "actuality" are also frequently used. What is actual, is that which we are experiencing. In this case, actuality is shaped very individually: Actuality depends on the observer. What exists as real, based on actuality, cannot be judged, since actuality cannot be separated from the observer. On the other hand, reality describes what exists independently of our observation.

In this strict sense, we do not know whether there is a reality and what constitutes this reality. In psychology, this insight leads to constructivism, namely the statement that we open up the world through our mental models and that these mental models are shaped by our entire personality. At the same time, these models are also an essential part of our personality. The question of reality and idealism (i.e. mental models, therefore ideas) is a question which dates back to Aristotle and Plato.

This question on reality and actuality achieved practical value with the second revolution in physics: The Theory of Relativity was followed by another theory, Quantum Mechanics. In the subatomic range, quanta behave as if they are influenced by the mere observation of human beings. Depending on the observation, we experience a different actuality. The question of reality still cannot be answered (so far?). Quanta without observation are only possibilities: The most famous example of this is the phenomenon of wave-particle duality in Quantum Mechanics. Prior to the advent of Quantum Mechanics, properties of quanta discovered, assigned to completely different types of objects dependent on experiments (i.e. particles and waves).

Since Popper [2], there has been the idea that theories cannot be verified, but only falsified. In other words, it is not possible to show that a theory "always" describes an actuality "correctly". This would require us having a God-like, all-embracing understanding of the world and reality, to be able to judge whether a theory describes reality. Our understanding of the world is always temporary, so we can state that a theory can not be wrong, as long as the actuality is adequately described.

It is also necessary that theories do not lead to contradictions with other confirmed theories, thus stimulating the co-development of theory and practice. Theory and practice together provide preliminary insights and knowledge. And another, even more important principle, that is not always considered, is that if several theories exist, it is always necessary to select the one that requires the fewest assumptions or parameters. We can also refer to "Ockham's razor."

This principle is fundamental in natural sciences. In the discovery of the above-mentioned theories, the theories of relativity and of Quantum Mechanics, this was a guiding principle. If we follow current discussions on particle physics or cosmology, we could conclude unfortunately that this principle no longer has relevance. The current unsatisfactory situation, especially in particle physics and cosmology, is "crying out" desperately for simplification [3].

Theories are essentially based on ideas and related models. Ideas are more or less structured thoughts that belong to a topic. In the case of the Theory of special Relativity, the topic is "the behavior of the speed of light in differently uniformly moving systems" and the associated unusual behavior of the speed of light. The idea was "if the speed of light remains constant and the definition of velocity is not questioned in principle, lengths or times must change as a function of uniformly moving systems". In hindsight, a fairly simple idea.

In this way, the centuries-old pattern of thought of the immutability of space and time was overthrown. Based on this central idea, the operationalization of the equal treatment of space and time was born: Time was introduced as a so-called fourth dimension in addition to the three spatial dimensions. This is a very important insight and an elegant mathematical representation, but it also had the future side effect of mathematical conceptualization being given priority with respect to "playing" with physical phenomena [3]. Precisely formulated, this means: At first there is mathematics, and only then do we think about theory and practice. As explained in the section "Regulation of Complexity Through the Formation of Models and Intuition", this clashes with Einstein's method of obtaining insight.

The models contained in theories depict parts of actuality and are valid under certain premises, which can be included as a set of axioms in the models. The models answer questions raised by the theory, that is, in the Theory of Special Relativity: "What are the consequences if the speed of light in uniformly moving systems remains the same?" The operationalization of the models often leads to a new understanding of terms and objects: In the theory of special relativity these include, among others, the terms "inertial system", "time", "space", and derived statements, such as the famous formula for the equivalence of energy and mass ($E=mc^2$).

It is mostly unspoken knowledge that models are based on principles and values which, for their part, are not explicit, e.g. are contained as axioms within the models. In the case of the Theory of Special Relativity, this could be the principle that time and space have the same behavior for all lengths of time and time scales and that space and time themselves have no particular microscopic structure. (This is a principle which has recently been increasingly questioned from a theoretical point of view, but there are no experimental indications to date). An associated value is belief in the validity of the macroscopic principles of time and space in any given world context.

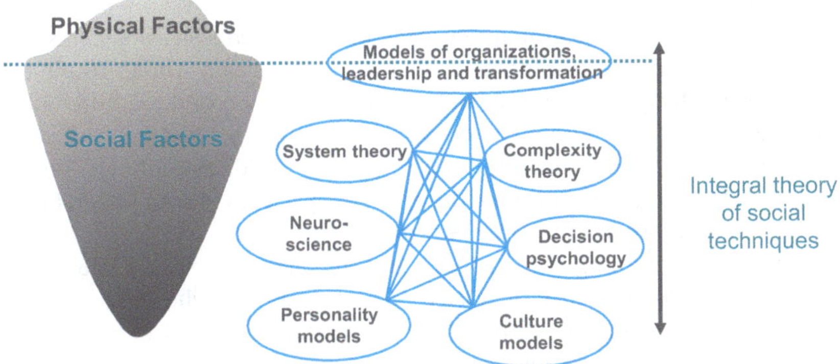

Fig. 7.1 Integral theory of social techniques

On the basis of the above understanding of "theory and practice", we present, in this book, an integral theory of social techniques for complex systems. Figure 7.1 illustrates how we adapt a variety of theories and models from natural and social sciences to the requirements of organizations, leadership and transformation, and connect them with a network of theories and models. The goal is to provide models for organizations, leadership and transformation, with which we can look behind the surface of social mechanisms to decipher their network of interactions. As described above, with the example of the Theory of Relativity, we repeatedly apply the fundamental ideas of the Dilts Pyramid to the formation of a single theory, as well as to that of the theory network. We do this from the context of the theories, via their values and basic assumptions to their goals, by illuminating all their relevant aspects in this book – always based on the insight that these aspects are limited and temporary.

Abstraction, a Term with Multiple Perspectives

In the glossary, abstraction is defined as a form of bundling, or formation of categories. We use the words "abstraction" or "abstract" in this book in different places and suggest that the levels in the Dilts Pyramid represent levels of abstraction. In this case, we associate the concept of abstraction with a special competence, meta-competency. Reiss Motives are regarded as categories, thus abstractions, of motivational schemes.

We also characterize the language of people with NT- or NF- preferences as abstract, and those of people with SP- or SJ- preference as concrete. Last but not least, via MBTI, we combine abstraction, with intuition and bias. In the diamond model, one of the axes, degree of abstraction, is associated with the term complexity. Therefore abstraction, including through intuition, is a form of complexity regulation. We also have a feeling of what "abstract" is: For most of us, mathematics is abstract. Here, abstraction has something to do with the ability to form models. We associate the distance to the tangible world with "abstract". The step towards "theory-practice", or "gray-colored" associations, is not far away in this case.

Confusing?!

When we speak of an integral theory, we mean that such confusions, or similar confusions, are resolved in our mind and thus experienced as a whole. In the following paragraphs we will try to offer some help on the integration of these different perspectives of the concept of abstraction. We emphasize that this can only be an aid: The authors are expressing their (integral) understanding by "disassembling" into different perspectives. The integration of these different perspectives into an integral theory can, in our opinion, only occur in the mental world of the reader.

Let us start by recalling some aspects of the Theory of Special Relativity. After being introduced to the theory of relativity, those students who have an aptitude for natural sciences, have no great difficulty in reenacting the Theory of Special Relativity from experimental facts and knowledge of the above-mentioned basic assumptions. Learning the theory for the first time may be considered alien, yet it becomes easier each time one approaches it. There then comes the time when the Theory of Special Relativity is self-evident; it has lost its "abstractness" for the student. Despite the fact that the theory is far removed from everyday life, and therefore has a certain degree of abstraction, once we are familiarized with it, we then start to view this abstraction as natural.

Let us for a moment, slip back to Einstein's time, before the formulation of the Theory of Special Relativity. What would our situation be? No one would be able to familiarize us with a solution to the problem, or introduce the solution. Purely analytical intelligence (S-preference) would therefore reach its limits, especially in cases of innovative or complex problems. It is true that a more abstract, i.e. generalized, representation could be derived from the details using analytical methods, but the result is implicitly defined by the individual steps of the analysis. It is quite rare in this way, to discover something new.

In other words, at one time or other, someone must have followed "the new way". An additional mental process is therefore necessary for solving unknown complex problems. It is necessary to step back from practice, and from the details of the problem, and to look at the practice from the outside. The means by which this step can be accomplished is intuition (N-preference). In this sense, intuition enables abstraction. This step is particularly difficult when it is carried out for the first time, and it is particularly very difficult when it is carried out for the first time by someone in this world. We assume that people have different ways of jumping from the level of detail to a more abstract level.

When someone carries out this step for the first time, the predominance of preference is most obvious. We assume that the S-N dimension in the MBTI expresses this preference. As we have already explained, the difference between S-preference and N-preference is most evident when an innovative context (e.g. an innovative project) exists. However, it is not only in creative processes that basic mechanisms of intuition work, but "always". The tendency to generalization, as a form of abstraction, is part of human nature. As we have seen, we all tend to form rules of thumb, however, experience has shown that people with an N-preference prefer this way of abstraction and also apply it often. Associated with this, is the preference for abstract language expressed by the use of generalizations and formation of categories.

To help to clarify this, we can view S-preference people as those who consciously analyze details (System 2) into categories or a generalized perspective. N-preference people often do this without deliberate intervention and unconsciously (System 1). They are mostly only intuitively aware of the result. Neurologically, only a limited amount of information can be consciously integrated into a state of consciousness, and each state of consciousness corresponds to a sequential step, such as a sequential step in an analysis. Unconsciously, however, parallel processing of a much larger amount of information is possible. These neuroscientific theses make it plausible as to why it is more likely for N-preference people to discover something completely new.

From intuition back to abstraction! The ability to evaluate details of already known types of problems at a more abstract level or to use them in comparable situations is often the result of training. Just think of people who have received vocational training first, and then academic training later. As a result of studying abstract concepts and models of known problems, those with vocational training are often unintelligible to others who have not trained in this field.

This is true, even if people in both groups have an S-preference. The Collective Mind Scheme with its three tiers of abstraction helps to bridge these different mental abstractions and functions. Individuals with a "concrete" preference (S-preference) help those with an "abstract" preference (N-preference) by asking questions and thus making the results of intuition tangible. This process of illuminating the different levels of abstraction requires patience and trust from all those involved. The result of this illumination is something new, which can be understood by everyone, regardless of mental preference. The result is comprehensible and reusable.

Because, compared to S-preference people, the preference of releasing or ignoring facts and details is higher for N-preference people, so too is the risk of bias. "Letting go of details" and formation of abstract perspectives also has advantages: It creates a prerequisite for regulating complexity. However, this can only occur when this step to a higher level of abstraction does not lead to communication problems. Because, as referred to above, during conversation, the big picture is not recognizable at the beginning for S-preference participants, whilst N-preference participants do not advance the details and facts. In some cases, the step to a higher level of abstraction is too quick and too premature, resulting in bias from all participants.

"Letting go of details" is associated with a higher openness to new things, according to John Cleese, a prerequisite for an Open Mode of creativity. Or alternately, "sticking to details" leads to a Closed Mode of creativity.

Meta-competency emerges when people are able to mentally escape from their own context, i.e. to perceive themselves and their surroundings from the outside. For us, this is made possible by the ability to abstract: A system element is viewed from the perspective of a system, and this system is viewed from the perspective of a larger system. A perspective of "systems of systems" emerges. This transition from an element to a system, to the "system of systems" … is defined as the degree of abstraction in the diamond model, which is one of the reasons for the growth of complexity.

Meta-competency has fully developed when the ability for sovereign change between the tiers of abstraction is reached. This means that a sovereign person can switch between details and the big picture, between experience and intuition, and between practice and theory. This prevents bias and models the New.

7.2 Fundamentals Complexity Classes

In the section "En Route to Complexity", we follow the Santa Fe Institute [4] in our understanding of complexity, according to which, natural, technical and social systems can show complex behavior. Complexity is not a "fixed property" of a system: A system is not "non-complex" or "complex", but the same system can show "non-complex" or "complex" behavior depending on certain system parameters. Systems with few interacting elements, as well as systems with many interacting elements, can both have complex behavior. Systems with many interacting elements, however, have a greater affinity to show complexity. Complexity can produce new emergent structures: The whole is more than the sum of its parts.

According to complexity research, there are four different complexity classes: the Wolfram classes [5]. In Table 7.1, we have listed these complexity classes from the point of view of system theory, in which we use furthermore the terms "simple, complicated, complex and chaotic" introduced by Snowden [6] with the Cynefin Framework in the management. As we are very well aware that these terms can be misleading, Table 7.1 illustrates the complexity classes by using examples from technology, natural sciences and organizations. In the description, there are two terms that are often repeated, trajectories and attractors: Trajectories refer to curves in a space that is spanned by the parameters describing the system. The parameters in natural-technical systems of this so-called phase space are often space and velocity or momentum.

Trajectories refer to the dynamics of the system. The technique of trajectories can also be used in social systems to make team dynamics visible during team coaching. Depending on the goal of the coaching, a variety of social phase spaces can be used: In [7] we illustrated the learning behavior of a team using our learning model: The three system parameters, according to which social phase space is spanned are: Learning/training culture, specialization of the language, and degree of abstraction of the language. Another three-dimensional phase space alternative is to display behavior (resonant communication, dissonant communication, no communication) against the S-N dimension (detail-oriented communication and big-picture oriented communication) and against different target directions. Based on the temporal course of the states in the phase space, the team can very quickly see dynamics that would otherwise be completely hidden.

If states are repeatedly reached over time, we talk of attractors. Thus, e.g. during team communication, a target may be repeatedly touched upon with the same degree of detail, even if additional new ideas occur during the course of the communication. In this case, there is a so-called point attractor. Or to take another example, there are discussions, in which (almost) identical states are gone through, without any substantial change in communication. In such cases, we speak of periodic attractors.

In addition, the emergent organizational structures, which can develop in complex systems by means of self-organization, are special attractors. The Collective Mind state of a team is a special attractor with a new mental team alignment.

Table 7.1 Wolfram classes and their manifestations

Wolfram class/ classification	Perspective of system theory	Manifestations in technology and natural sciences	Manifestations in organizations
1 Simple system with stable order	Systems with very simple trajectories i.e. point attractors, are reached quickly. The systems are non-sensitive to initial states.	Simple logic circuits, many mechanical devices, (but note: not all).	A rigid hierarchical organization always reacts to all market conditions in the same manner. In meetings, information from participants is not integrated, but an agenda is rigorously pursued.
2 Systems with periodic patterns: Complicated, ordered systems	Systems with point attractor trajectories and which, in the dimension of time, are periodic attractors. The systems are sensitive to initial states.	Smartphones are usually complicated. However, if there are interactions between apps, complexity may occur. The design idea of independent apps is a good guarantor for complicated behavior.	An organization has some stereotypical behavior; in some cases, it looks as if new behavior exists, but the pattern repeats itself in certain cycles. In meetings, information is repeated on a regular basis, so that no new findings are obtained. Superficially, there is the impression that variety exists.
3 Chaotic Systems	Systems, which over time, have chaotic trajectories and are extremely sensitive to initial states. There are no conditions for emergent structures.	Electricity or railway networks show high levels of back-coupled interconnection. In principle, these systems can become chaotic.	A team is behind schedule. Comments by the project leader become the straw to break the camel's back. Other stakeholders are activated by the team members. The situation escalates and cannot be stabilized over time.
4 Complex Systems with emergence	Systems that have simple attractors and show high sensitivity to initial states. Local structures of order are formed and many different states are generated. Emerging structures arise.	Laser: A dominant light wave forms and "enslaves" all other light waves to this frequency. A light beam of extremely high intensity is generated Thermal Convection: From initially unstructured microscopic flows, macroscopic spatial flow patterns form which can be stable or unstable over time,	A team shows a high level of mental interconnectedness. There are mental attractors like the Collective Mind. The Collective Mind is created after a variety of information has been processed at the beginning, to finally realize a dominant mental state of information in the team.

7.3 Fundamentals Dilts Pyramid

The bon mot; "Problems can never be solved in the same way of thinking from which they originated [translation by the authors]" is attributed to Albert Einstein. This is the main basic assumption of learning that underlies this book. It is also one of two reasons why we have chosen the Dilts Pyramid as a "master model" for this book: The Dilts Pyramid is a special model of levels of the mind that helps us to look at ways of thinking, not only from a different perspective, but also from perspectives that lie one or more levels above the resulting problem.

The Dilts Pyramid is a very simple, but very effective tool for resolving problems. The term "resolution" was used specifically here, rather than the term "solution": Problems are not only solved, but will be resolved by changing the perspective to a higher level. There is no "either…or" but instead they dissolve into a "both …and". In the section "Learning and Meta-Competency" in the chapter "Leadership in Complex Social Systems", we have already drawn a link between the Dilts Pyramid and Bateson's stages of learning. In this section of the appendix, we summarize the basic principles of our understanding of the Dilts Pyramid.

The Dilts Pyramid was introduced into NLP in the 1990s by Dilts, and was connected to the transformation work of belief systems[1] [8, 9]. It is assumed that Dilts was inspired by Maslow's hierarchy of needs, the Bateson stages of learning, and the famous "model of the logical types" by Russel and Whitehead [10].

Figure 7.2 shows the Dilts Pyramid with a brief description of the respective neurological levels. Sometimes the relationship to the model of the logical types shimmers through and logical levels are referred to. We do not, however, assume that the levels of the pyramid are specific features of the logical types according to Russel and Whitehead.

And we also do not assume that the neurological levels correspond to neurological correlates. The basic idea of the pyramid, read from bottom to top, is that the number of "elements" decreases from level to level: The "environment" level includes the many different contexts in which we live. To name just a few: project meetings, meetings with management, visiting the cinema with children, visiting parents or in-laws, etc.; we are always on the move in different contexts.

In different environments, we display different behaviors. The pyramid expresses the fact that while there are still many behaviors, there are still fewer behaviors than there are environments. And indeed, in different environments, we can observe ourselves having the same, or at least similar behavior. We display this behavior by accessing our abilities: Either to discuss, to give a lecture, or to help one of our children in their development.

The number of skills is already significantly lower than the associated number of behaviors and, of course, skills have an impact on the underlying level of behavior. This also applies to the other levels. Higher levels have an impact on underlying levels. But the

[1] We speak of a belief system, because beliefs that "fit" together, attach themselves to others, and so a system of beliefs arises.

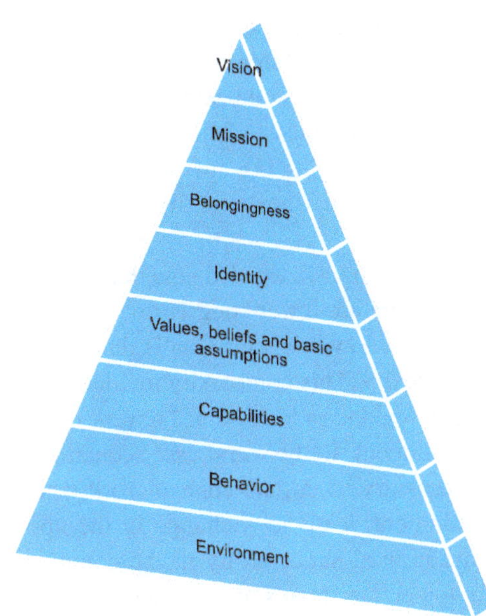

The level "**Vision**" points beyond the system and answers the question of why by where the system travels to.

The level "**Mission**" answers the question of why by the function which the system fulfils as part of the big picture.

The level "**Belongingness**" begins the endowment with meaning by assignment of the system to the big whole.

The level "**Identity**" contains the system kernel (temperament, motives) with a self-image, which determines the structure of the lead-memes.

The level "**Values, beliefs and basic assumptions**" consists of lead-memes gained by experience, which gives the capabilities guiding principles.

The level "**Capabilities**" represents strategies, on which the system behavior is based.

The level "**Behavior**" stands for perceived modes of behavior of a system in its context.

The level "**Environment**" describes the environment (context), with which the system interacts and acts accordingly.

Fig. 7.2 The Dilts Pyramid

reverse is also true. Because environment has an impact on our behavior, this also leads to a change in the abilities we use. Experiences are thus formed from "below" to "above". So much so, that at the level of values and basic assumptions through recurring experiences – by socialization – values and basic assumptions are formed.

One of the most obvious examples of such an occurence is the "health" value. It may well be that this value has only gained significance as part of an experienced disease. If this value, formed by experience, exists, it affects our abilities and subsequently our behavior, like a guideline: We will possibly avoid working overtime, and participate in more sports and spend more time with the family. Basic assumptions include principles, beliefs, heuristics and rules of thumb. The manifestation of the rule of thumb "customer orientation", i.e. "you stop everything when a customer calls", is probably the result of behavior in different environments, where such behavior has led to positive effects: The customer was friendly, he placed an order and the supervisor gave praised. As we have seen in the section "Regulation of Complexity Through the Formation of Models and Intuition", rules of thumb are mental tools that extract patterns from a variety of experiences.

They reduce complexity in our perception. The preference for extracting a specific rule of thumb is not determined solely by the levels of environment, behavior, and capabilities. The ability to adapt quickly to a new situation is also regulated by preference, which is pre-arranged by temperament and basic needs, namely the level of identity. A person who likes to act in a situation-oriented manner (preference Perceiving in the MBTI) will experience less negative stress when an unexpected call comes in, than a person who just wants to complete a task (preference Judging in the MBTI). If the temperament preference is also supported by an appropriate preference in basic needs or motives, this reinforces the formation of corresponding values and basic assumptions: A preferred basic need for emotional calm, control and orderliness, would probably resist forming the "customer orientation is when you stop everything if a customer calls" rule of thumb. Instead: "customers are the ones that keep you from your work" might be established.

For the levels "identity" and "values and basic assumptions", we also note that there are fewer elements, that is, fewer parameters, and thus there are fewer "degrees of freedom". This "fewer" simultaneously affects the "more" on the lower levels. This is the quintessence of what we understand by regulation of complexity. This complexity regulation can have both positive and negative effects. Complexity regulation hinders, if "fewer" no longer fits into "more and different" of the level environment. In the worst case scenario, the upper levels act as blockers. In this case, it is necessary to initiate change work which, taking into account the respective current environment, leads to a change in the upper levels. This is the reason why the Dilts Pyramid is used successfully in the context of change work. In doing so, change work can refer to persons, teams or organizations, i.e, to all social systems. For this reason, we have spoken about it more generally in the brief description Fig. 7.2 of "system".

Up to and including the level "identity", the level structure of the Dilts Pyramid is used almost uniformly in literature and practice. In its original form, the level "belongingness" condensed our threefold level structure "vision, mission and belongingness" into one level. These three levels have a transcendent dimension, which can also have a spiritual character, depending on their manifestations. These are the levels that serve as a model for giving sense and meaning to our being and actions: The present, in the form of the environment, in which we act with our personal preferences and needs, is adjoined to our past and future.

People, as well as organizations, develop a vision and a mission, and feel a sense of belonging to other systems. Belongingness for an individual, may mean belonging to a professional group, whereas for an organization this could mean belonging to an industry. Belongingness can also mean feeling a sense of belonging to life itself. In any case, this belongingness extends far beyond the system under consideration. An elaboration of the "Vision" and "Mission" levels is also carried out in this way. Both levels produce a meaningfulness in which they regard the tasks of the system in the light of another larger system and subsequently assign them a larger or superior purpose.

An environment-specific expression of all levels has been described elsewhere as a mind-set or inner attitude. We therefore speak of something being "environment specific", because the mind-set that a person has within the environment of a project team does not have to match the mind-set that this person has in a departmental meeting or when they are together with their family. It is likely that the mind-sets are similar, or at least equal, at the "identity" and "values and beliefs" levels. However, we also know that, depending on the environment, different temperament preferences or different weightings of motives are shown to the outside world. In such cases an "individual" is likely to "work" against their own strengths and basic needs. In the worst case scenario, when there are completely divergent upper levels of meaning for different environments, this results in unmanageable stress, which can lead to long-term illness.

Since a project is a temporary organization, it is also necessary to temporarily develop all neurological levels. The main leadership tool is the Collective Mind Scheme with consists of the three tiers: "Goal", "What" and "How". The Goal-tier corresponds to the "vision" level, the What-tier corresponds to the "mission" level. It is only at the How-tier, which contains the detailed objectives, where there is no direct correspondence to the level of "belongingness". However, it should also be mentioned that in the practical application of the Dilts Pyramid, detailed objectives are often built in, and this corresponds to the How-tier.

The Collective Mind Scheme, along with the top three levels of the Dilts Pyramid, clearly have the function of an order parameter hierarchy. Starting with the "identity" level, these levels are able to assume the roles of order, control and setting parameters: This is best seen at the "values and basic assumptions" level. Values which a team gives themselves voluntarily and amicably, act as an order parameter. Appreciation of the values of each individual team member by the team members and the surrounding organization, acts as a control parameter. Through appreciation, team members are able to flourish and contribute their strengths to the team.

However, values may also act as setting parameters, if certain values are specified by an organization to a team. These could be seen as very similar to order parameters. It should, however, be remembered that order parameters must come from the team itself and never from outside. Values that come from the outside can act as a straitjacket and can then prevent the creation of any added-value complexity. A company that wants to create a new, innovative product with a project, but at the same time needs the security of a fixed budget, is such an example.

The Dilts Pyramid represents the master model for our integral model. Figure 7.3 therefore, provides an overview of models used and those used in this book [7], assigned to the levels of the Dilts Pyramid:

Mind Viruses

A quotation from the book by Robert B. Dilts, Judith DeLozier, and Deborah Bacon Dilts [8] referring to a monkey experiment, illustrates nicely the power of unconscious "thought" (i.e. basic assumptions):

"The process begins with a cage containing five monkeys and a specially rigged water system. In the cage, experimenters hang a banana on a string and put a set of stairs under it. Before long, a monkey will go to the stairs and start to climb towards the banana. As soon as he or she touches the stairs, however, the water system is activated and sprays all of the monkeys with cold water. After a while, another monkey makes an attempt with the same result – all the monkeys are again sprayed with cold water.

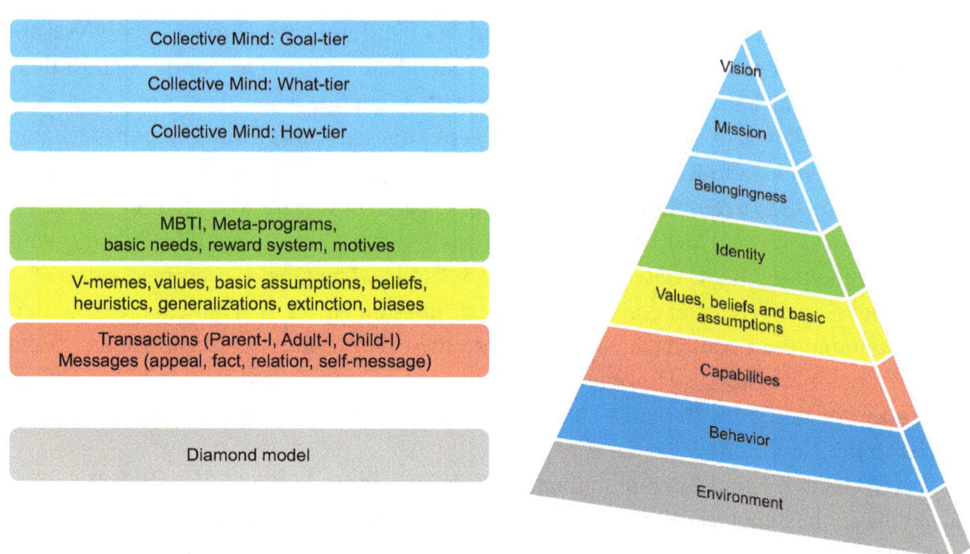

Fig. 7.3 Overview of the models used in this book

Pretty soon, the monkeys make the connection and when any monkey begins to head for the stairs, the other monkeys will prevent it, using physical force if necessary. Eventually, the monkeys learn to completely avoid the stairs.

The experimenters then deactivate the cold-water spray mechanism. The monkeys never discover this, however, because they continue to avoid the stairs. (The territory has changed, but their map of it has not! [Annotation by the authors: This is one of the key principles of NLP])

Things become even more interesting when one monkey is removed from the cage and replaced with a new one. The new monkey sees the banana and wants to climb the stairs. To his surprise, as he approaches the stairs, all of the other monkeys rush up, and restrain him. The more he struggles the more forceful the others become. After several attempts, he learns that if he tries to climb the stairs, he will be assaulted.

Later, another of the original five monkeys is removed and replaced with a new one. The newcomer goes to the stairs and is restrained by the group as well. The previous newcomer apparently joins the others with enthusiasm ('If I can't go on the stairs and get the banana, neither can you!'). The second replacement monkey also eventually learns that the stairs and the banana are 'taboo'. After this, a third original monkey is replaced with a new one. When the new one moves to the stairs, he or she is restrained by all of the others as well. Two of the four monkeys that prevent him have no idea why they are not permitted to climb the stairs (they have never been sprayed with cold water). They are mirroring how they have been treated.

Ultimately, the fourth and fifth original monkeys are replaced. All the monkeys that had been sprayed with cold water are now gone and the cold water has been turned off for a quite a while. Nevertheless, no monkey in the new group ever approaches the stairs. Why not? Because, that's the way it has always been."

Four Ears, Five Chairs

The following exercise supports the idea of "to learn to perceive" and to take meta-positions of perception. The exercise is based on the four-eared model of communication [11]. The participants in this exercise are asked to slip into different perspectives of a conversation and will be encouraged to formulate and articulate these perspectives. The exercise is designed for five roles and persons, respectively:

- The participants are requested to sit down in a circle with five chairs. Each participant will be assigned a role.
- The first participant (narrator, role 1) tells a story that has happened to him or her.
- The second participant relates the facts (role 2) that they themselves have perceived.
- The third summarizes the feelings (role 3) that they have heard.
- The fourth participant talks about the needs (role 4) they believe they have perceived from the story.
- The fifth talks about the solution options (role 5) which they have considered for this problem.
- Subsequently, the narrator decides with whom he felt most comfortable.
  Then the players move around one place clockwise, the roles remain associated with the original chairs.

7.4 Fundamentals MBTI

In this book we model people's personality through their individual manifestation of the Dilts Pyramid. In particular, we use the temperament model and the model of basic needs and the associated reward systems, simplified by using motives, for the modeling of the logical "identity" level (please refer to the appendices "Fundamentals Consistency Theory", "Fundamentals Reiss Motive Profile" and "Fundamentals Dilts Pyramid").

The temperament model is based on the assumption that every human being has a preferred way of reacting to, and acting in, their environment. Ideally, a comprehensive theory, which describes cognitive processes and an individual's associated actions within their environment, according to "first principles", is preferable to be used for this purpose. Unfortunately, to date, no such theory exists. There are two possibilities: Use general models describing people's preferred and typical modes of action, or elaborate validated tests by means of statistical methods, which enable statements to be made about designated action patterns in specific contexts.

A general model has the advantage that it describes certain action patterns, but it is also general; so it cannot always be used reliably for prognostic purposes. Statistically validated tests, on the other hand, allow a more accurate prediction of defined requirements profiles, but it is difficult to generalize the specific results. Current research is only too aware of this unsatisfactory situation, although discussion is sometimes emotional or related to commercial considerations. We have decided to model the behavior described in this book, as far as temperament is concerned, using MBTI typology (Myers Briggs Type Indicator Typology). The reasons are: Firstly, with respect to temperament, typical patterns of behavior are sufficient to address project situations. A more detailed description would not be useful.

Secondly, the four main dimensions of the MBTI can be consistently assigned to the Big Five personality model [7, 12], so that there is a meaningful and validated connection to the current standard inventory in personality psychology. It should be noted that although MBTI typology is based on C.G. Jung, there are, however, substantial differences in approach. Jung's model is cognitive-driven (thinking processes are in the foreground), MBTI, on the other hand, is action-driven (observed behavior is in the foreground). In addition, we do not consider MBTI as a dichotomous (either-or) scale, but a continuous scale, located between two poles, as is nowadays common in the literature. MBTI includes the term 'typology' in its name, which corresponds to a dichotomous scale. Since at least the 1980s, Keirsey [13, 14, 15] has used a typical reading of dimensions with continuous scales, each with two poles, in personality research. Criticism still exists, but is therefore not justified, because the Golden Profiler of Personality (GPOP), which is proposed as an alternative, corresponds almost exactly to MBTI according to Keirsey [16].

The following paragraph is a summary update from [7] and [17]. We will not go into the description of the Big-Five personality model here and refer again to point [7].

MBTI typology is thoroughly elaborated [18, 13, 12] and in three of its four dimensions it relates back to C. G. Jung [19], as far back as 1921. In [19] (blurb) the following statement is made [translation by the authors]: "Jung's type theory is not a system to schematize and statistically diagnose people, but a way to give orientation in the individual development process that is always in flux, to find help to understand oneself, but also to understand your fellow human beings." It is in this sense that we also understand MBTI typology and its application.

The modeling of a person's temperament is carried out according to MBTI typology through four pairs of temperamental "poles": E-I, (extraversion versus introversion), S-N (sensing versus intuition), T-F (thinking versus feeling), J-P (Judging versus Perceiving). Table 7.2 summarizes these dimensions and provides clues for an initial understanding.

A very detailed description of MBTI types and their meaning can be found in Keirsey's books [13, 14]. In the following, we will discuss the characteristics of the temperament dimensions, typical behavioral patterns for each dimension and stress-triggering factors.

Let's start with the first dimension, social interaction. If a person takes their energy from contact with other people, they are extraverted (extroverted, both terms are used), and their state is closer to "E" than "I" on the "E-I" scale, i.e. they will be labeled with an "E". An introverted person, on the other hand, will lose energy in contact with other people, and is therefore more self-sufficient, and is closer to "I" than "E" on the "E-I" scale. This is marked with an "I". Roughly speaking, an introverted person thinks first and then talks, however an extraverted person thinks while they are talking. In respect to this scale, we can also speak of people's social orientation. Extraverted temperaments love to contribute to conversations to maintain personal exchange. They cultivate personal contacts

Table 7.2 Temperament dimensions according to MBTI

Extraversion (E)	Social Interaction (Source of energy, contacting)	**Introversion (I)**
....outwards (Sociability, interest in external events, acting, outside world, energy intake in contact with others, easygoing, self-expression, speaker)	Focus of attention...	...inwards (Being alone, interest in inner processes, thinking, reflecting, writing, reserved, inner world, energy out of itself, listener)
Sensing (S)	Exploitation of information and problem solving	**Intuition (N)**
...facts and concrete (Concrete, feasible, present, facts, objective, repetition, practical, reasonable, facts, experience)	Concentration on....	...opportunities and patterns (Opportunities, future, abstract, intuition, pattern, meaning, conceptual, speculative, theories, inspiration)
Thinking (T)	Decision-making (Information assessment, decisions)	**Feeling (F)**
...analysis and objectivity (Analysis, causal principle, rules, equal treatment, logical arguments, principles, factual, objective, logic)	Decision-making according to...	...values and feeling (Empathy, orientation on personal values, subjective, heart, sympathy, personal conviction, consider individuality)
Judging (J)	Structural requirements to environment... (attitude to life)	**Perceiving (P)**
...organized and planned, judging (Planning, considerate, controlling, regulating, systematically acting; assigning, doing and finishing things, avoiding stressful situations)	Prefers...	... spontaneous and flexible, perceptive (Spontaneous, flexible, adaptive, understanding and changing, keeping things open and flexible, energy in stressful situations)

and like to have many interesting and communicative people around them. Introverted temperaments, on the other hand, think before they express anything. There may be a pause between enquiry and response. The stress level for E-temperaments is high when they have to work alone on a project or are forced to wait for work results from others. For I-temperaments, increased stress occurs when they have to work on group projects, when they are frequently disturbed, are forced to act or to give (spontaneous) presentations.

The next underestimated, and even more significant, distinctive features of MBTI are; the way in which information is received and elaborated, information exploitation and problem solving: S types are more interested in numbers, data, facts, i.e. concrete things.

N types prefer to think intuitively and holistically, i.e. in big picture concepts. S-types are therefore called "fact-oriented", N-types "intuitive". Since the common perception and elaboration of information and knowledge as well as their interpretation is essential for co-operation between people, particular attention should be paid to these different ways of information exploitation.

For example, many scientists are N-types, but so too are some politicians or writers. This group of people does not focus on elaborating concepts and structures in detail, but rather on giving ideas and impulses, or shaping them, without having to go into detail. So it is mainly S-types, who have a special interest in detail. They are happy to work in organizations such as administrations or companies and prefer pragmatic, concrete objectives and tasks. They ensure stability and consistency.

It should be noted that N- and S- types understand "structure" differently. S-types place emphasis on small-scale details, clear guidelines and a defined context. For N-types, a conception, a picture or a goal representation is sufficient as a structure. These two understandings of structure are therefore fundamentally different. It is also interesting to note that empirical studies in the US have shown that about 80% of all people there are S-types. We see this result as essentially transferable to other countries. Because of their numerical inferiority, N-types often face a lack of acceptance.

S-temperaments therefore tend to relate and demand lots of concrete facts (the "How"), whereas N-temperaments think in abstract concepts and usually in larger, sometimes discontinuous contexts, and express themselves accordingly, i.e. expressing fewer statements on the concrete "How". S-temperaments are under stress when, for example, they work with abstract material the context of which they do not know, or when the requirements in a project are too indeterminate, established products or methods are not appreciated or even ignored and it is expected that they spontaneously develop new methods. N-temperaments are under stress when they have to work with detailed material, right at the beginning of a project; requirements are too determined; and innovation or changes are either not appreciated or ignored.

Another distinctive feature, namely the way a person makes decisions, is expressed by the pair T-F: Rational ("T" like thinking) or by relationships ("F" like feeling). A person who makes rational decisions in a group will focus less on soft factors such as climate, well-being or bad feelings, and instead try to argue on the basis of logical processes and content; with this person being perceived as both sober and cold. A person who takes into account the feelings of others in the decision-making process, communicates accordingly and pays attention to mood and climate, will sometimes, instead of trying to convince, try to persuade. Some of the statements made by these people will therefore appear illogical for T-types.

So how do we recognize F- and T-temperaments? F-temperaments very much like harmonious interactions, so that they can approach project team members and connect with them. T-temperaments, on the other hand, desire clear facts and statements to extend their knowledge. They appreciate and benefit from the experience of other project team members on specific topics. T-temperaments are under stress when competence is lacking, there is no objectivity or project team members ignore logic. F-temperaments, on the other hand, are under stress when there is an absence of cooperation and harmony, when they are asked for criticism or the feelings of others are ignored.

The fourth and final element in the MBTI structure indicates whether a person prefers to act by judging (methodical) (letter "J" like judging) or perceptually (letter "P" like perceiving), i.e. acting based on a situation. Hidden here is another potential for conflict: Perceiving-oriented people follow less of a fixed timetable, on the other hand, judging people lack spontaneity. For example, a perceiving-oriented project leader will most likely be considered "chaotic" and unreliable by judging-oriented project team members. On the other hand, perceiving-oriented project managers may see judging-oriented project team members as "picky". Perceiving-oriented people are situation-oriented, judging-oriented people are result-oriented.

This function describes fitting into the environment. J- and P-temperaments can be recognized as follows: J-temperaments are those who contribute to the team, without losing sight of the goal. To this end, they are also willing to abandon, for example, a facet of a project solution, which in their opinion does not contribute to the overall picture of the solution. P-temperaments are spontaneous and flexible, but tend to immediately take into account new emerging ideas, aspects or emotions. J-temperaments are under stress when everything is totally undecided and unstructured, if a result is not in sight, if they are asked to abstain from judgment and if they have to change their plans. P-temperaments, on the other hand, are under stress when everything is planned tightly, a result is quickly available or should be present, if they are asked to make a spontaneous judgment and they are unable to change their plans.

By combining these four "complement-pairs", 16 (simplified) MBTI-types are created and each person can be modeled according to one of the 16 MBTI-types. We would like to point out again that the temperaments are on a continuous scale between the two poles of each of the four scales, and not on the poles themselves. Of course, an introverted person will also have extraverted moments, an intuitively thinking person can think in an amazingly detailed way, an analytical temperament can appear warm, and a decision-maker can "turn a blind eye". But: The respective individuals will have to activate more energy in such cases, as they have moved away from their default state, so are forced to work against their strengths.

According to Keirsey, the temperaments can be divided into so-called function pairs: NT, NF, SJ and SP. A corresponding assignment of the temperaments to the pairs of functions is as follows:

- Idealists are labeled as temperament groups _NF_, i.e. ENFJ, ENFP, INFJ, INFP
- Guardians are labeled as temperament groups _S_J, i.e. ISTJ, ESTJ, ISFJ, ESFJ
- Rationals are labeled as temperament groups _NT_, that is, ENTJ, ENTP, INTJ, INTP
- Artisans are labeled as temperament groups _S_P, thus ESFP, ESTP, ISFP, ISTP

Figure 7.4 shows that temperament orientation N is combined with the temperament dimensions F-T: These are the temperaments of abstract language. Temperament orientation S is combined with the temperament dimensions J and P, the types with concrete language.

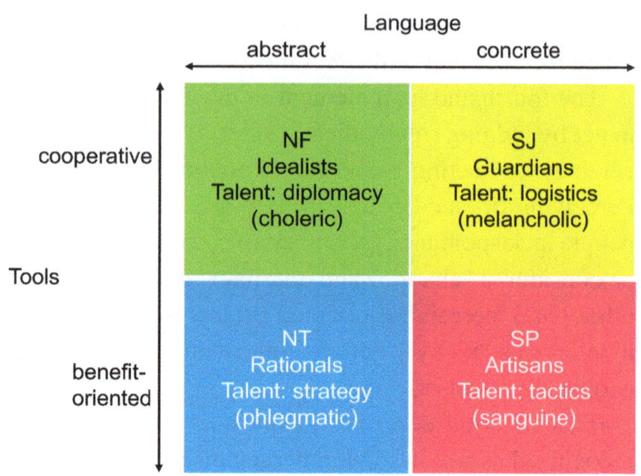

Fig. 7.4 Function pairs according to Keirsey (color coding corresponds to that of Keirsey, but does not correspond to that of Spiral Dynamics: Idealists: green, Guardians: yellow, Rationals: blue, Artisans: red)

With respect to cooperative behavior, these are the idealists and the guardians, who use their tools cooperatively, i.e. they are guided by what *the team* considers meaningful in the cooperation. The rationals and the artisans are, however, less cooperative in their selection of tools, i.e. they orient themselves to the task and what *they regard as suitable tools for the task.* (Figure 7.4 contains the assignment of temperaments according to Galen (approx. 200 AD) made by Keirsey himself.).

In practice, function pairs have the following characteristics:

NT are logical, imaginative types with high level analytical ability, who need self-controlled competence for their respective project domain. They prefer communication where they do not need to deal with emotional needs. It is important for them to develop the whole.

NF are enthusiastic, understanding types, who are aware of their abilities. Working in a team must contain potential for development; they look at the big picture and are sensitive towards other people. In addition, they think on an interdisciplinary level in the team and take into account the future orientation of a project.

SJ are practical, sober types with high level decision-making competence. They demand benefit and project planning in detail, as well as the presentation and measurability of effectiveness and efficiency at each individual project step. They need all relevant facts for decision-making.

SP are practical types with high manual competency. They perfect their practical skills, whether in the manual or in the artistic field. They need an environment in which they can "work off", and use their abilities knowing that they are appreciated.

For the sake of completeness, we should mention that the first so-called meta-programs of NLP [20, 21] were derived from MBTI. Meta-programs are personality-specific perception filters. The most important are facets (these are sub-dimensions) of the MBTI dimensions. They thus detail the MBTI dimensions. In the course of the further development of the meta-programs, however, meta-programs were also added, which cannot be presented as facets of the MBTI-dimensions. Examples of meta-programs are "complement-pairs" proactive-reactive, internal-external, self-others, which are assigned to the E-I dimension or the detailed-holistic "complement-pair" assigned to the S-N dimension. In NLP, meta-programs are combined with specific manifestations of language and body language.

The S-N Dimension and "Building a Small Home"

The S-N dimension is, according to our experience, very important for carrying out complex and innovative projects. Therefore, it is useful to stimulate self-reflection through a "practical" exercise. The following workshop exercise, of approximately 30 minutes, with subsequent reflection in a plenary session, supports this. In preparation for the workshop Tangram's paper should be printed out and the geometrical figures cut out [22]. Each workshop participant will receive an envelope containing these geometric figures. The workshop participants then "build a small house" with the following instructions:

Please build a house using the paper shapes. There are no limits to your imagination.... Describe the house so that another person can reconstruct it. Make a sketch. Cover your house.

Find a partner to build the house according to your instructions (and vice versa). Use the paper shapes, and the sketch will enable you to construct the house quickly. Of course, your partner must not have seen your house prior to this. Compare the "original" and "replica". Talk to your partner about your experience.

In the discussion that should follows in the plenary session, it will soon become obvious that various possible descriptions of the small houses exist, which can be assigned easily to the three levels of the Collective Mind target hierarchy. In this way, participants immediately learn that their personality is directly reflected in their actions. At the end, participants will be able to share their descriptions on the whiteboard, which is divided into 3 sections for this purpose. This results in a visualization of the team structure for the S-N dimension. Depending on the project or the team, the moderator will encourage discussion on the consequences of goal definition in projects, concept elaboration and requirements engineering.

MBTI Sociogram and MBTI House

In order to get to know the self-reflection of your own personality and the effects of interactions between different personalities in the team, it is very helpful to apply an MBTI sociogram. For this purpose, a reduced questionnaire of e.g. three questions per personality dimension [13, 23] is projected onto a wall, and workshop participants are asked, by referring to these questions, to assign themselves to a preference.

It is important to point out to participants that this initial self-reflection is not about accuracy of personality assessment, but about the openness to perceive this self-reflectively for themselves and for others. It is important to remind participants that they should answer the questions with respect to their private life and answer as quickly as possible. The fun of the team experience is the main focus. After participants have assigned themselves to a preference and written it down on a moderator card, the moderator places two cards, labeled with MBTI dimension personality poles, at a sufficient distance apart on the floor of the room.

	.ST.	.SF.	.NF.	.NT.
I..J	ISTJ Meticulous Analyser (Inspector) sound, settled, objective	ISFJ Reliable Guardian (Protector) friendly, conscientious, liable	INFJ Sober Teacher (Counselor) patient, interested, dedicated	INTJ Intuitive Thinker (Mastermind) goal-oriented, critical, independent
I..P	ISTP Practical Researcher (Crafter) cool, analytical, humourous	ISFP Loyal Idealist (Composer) friendly, sensible, modest	INFP Reflective Idealist (Healer) reflective, literate, interested	INTP Theorist (Architect) perceptive, analytical, reserved
E..P	ESTP Practitioner (Promoter) functional, direct, content	ESFP Problem Solver (Performer) open-minded, friendly, co-operative	ENFP Judge of Character (Champion) witty, curious, self-confident	ENTP Inventive Initiator (Inventor) quick, interested, multi-skilled
E..J	ESTJ Organisor (Supervisor) prudent, functional, aware of own power	ESFJ Mediator (Provider) popular, conscientious, active	ENFJ Optimist (Teacher) interested, harmonious, liable	ENTJ Successful Leader (Field Marshall) eloquent, interested, self-confident

Fig. 7.5 MBTI house

The participants are then asked to position themselves around the room according to each dimension. After positioning themselves around the room, the moderator then asks the participants about their evaluations and encourages discussion on the consequences for work on the project. After all four dimensions have been "played out" in this way; participants have formed their first explicit mental map of their own personality traits and of those of the other team members.

If all participants agree, this mental map will be reinforced again by visualizing the so-called MBTI team house. For this purpose, participants receive stickers with which to position themselves in the MBTI house (see Fig. 7.5). In our experience, this strengthens team-building via self-referentiality and encourages further discussions on team-setting and -dynamics. Later on, these insights are mentally transferred to a more comprehensive self-reflection and discussion of the team interaction patterns by working on values and basic assumptions with "personality flowers" (see section "Fundamentals of Leadership").

7.5 Fundamentals Spiral Dynamics

People experience fundamental developments when fundamental beliefs about the world and themselves change. These clusters of basic beliefs include thoughts, emotions, values, and other beliefs, and are referred to as "levels of existence" by Clare W. Graves. Graves, an American, was an organizational psychologist and a contemporary of Abraham Maslow.

Between 1958 and 1986 Graves empirically researched the manifestation of beliefs in healthy adults. He empirically detected seven "levels of existence", which build upon each other and represent human development like an "oscillating spiral". During initial single studies, an eighth level also became evident. According to Graves, the evolutionary spiral oscillates by meandering between the levels, each of which emphasizes the development of an individual (self-expression) or collective (self-devotion).

"Levels of existence" develop, because people have to adapt to the concrete living conditions of their environment. "Levels of existence" are also reflected collectively in artifacts, cultural agreements such as language, as well as in values, norms and organizational forms. Each higher level includes the level below, and can provide more complex answers to higher complex needs of life. In each "level of existence", an individual, or group of people, responds adequately to the needs of the environment, but also creates new problems that can only be integrated and solved at the next level. This creates the need for the development of individuals and systems, as well as organizations.

Two of Graves' students and colleagues, Beck and Cowan, subsequently introduced a color code for each "level of existence" into the system [24]. They also connected the term "meme" to the "levels of existence", a terminus of the cultural evolution theory [25]. This launched the term "meme" in parallel to the term "gene", to explain the distribution of psychosocial patterns in society. Beck and Cowan published their Graves-based interpretations under the heading "Spiral Dynamics" and then named the levels v-memes (value-meme).

Within a person, the "spiral" of v-memes grows steadily with their life history, whereby one or two v-memes are centrally active. However, the lower layers of the "levels of existence" can be activated again when the environment stimulates them. For example, many people in large groups at soccer matches and rock concerts experience a clan-like connection, which usually only exists in traditional cultures.

Here we describe the "levels of existence" related to [24, 21, 26]. See also Fig. 7.6:

Level 1 beige: The survival-oriented stage of existence: archaic-instinctive.

V-meme: The world is a hostile environment that dictates physical demands, drives and instinct. Behavior is based on basic needs and **self-expression**, the instinct to survive.

This v-meme occurs in our society only in extreme situations (war, disasters).

Level 2 purple: The animistic or clan-related level of existence: clan, security-oriented, self-sacrificing.

V-meme: The world is menacing and full of mysterious powers and spirits, which must be appeased.

Life is characterized by **self-devotion** for the clan and its rules, the elders and the ancestors. Individuals belong to a community, and pass on the rituals of the clan, revere the spirits of the clan and find their places in the community. Clan-like cultures define ritual transitions that young people and adults have to pass through to reach the next stage within the community. In modern industrial society, there is often a dislike for the concepts and rituals that originate from this v-meme. Since they are viewed as unfashionable for modern management and society. A complete loss of symbols and actions from the purple level, however, leads to what is frequently lamented as "coldness" in large company

TURQUOISE	Holistic spaces	Nourish the spiritual affinity of togetherness
YELLOW	Integral Meshworks	Do what is necessary for the good of all, establish integration
GREEN	Values collectives	Protect minorities, seek collective
ORANGE	Modern states	Act optimistically, strive for top position
BLUE	Old nations, empires	Find meaning, respect structure and discipline
RED	Feudal empires	Demonstrate power and force, courage, vitality
PURPLE	Ethnic identity	Maintain tradition and myths, belonging
BEIGE	Survival as an individual	Secure elementary needs

Fig. 7.6 Spiral dynamics: the DNA of culture and the four basic needs

structures. Thus, the backlash grows, i.e. the search for rituals, community and belonging. Outdoor training is a modern expression of this.

Level 3 red: The self-determined, "heroic-life-without-consideration-of-others" level of existence: egocentric, hedonistic.

V-meme: The world is like a jungle, where the strong and the tough rule and the weak serve. Nature is regarded as an enemy that has to be defeated.

What counts is who wins the fight, no matter at what cost: Others have no importance; the only important thing is self-survival in the story of mankind forever.

At this stage, we experience **self-expression** in its pure energetic form without any rules or remorse. It is about the individual who stands out from the crowd. Be it Robin Hood or Attila the Hun, there are both villains and heroes at this level. However, it is the individual deeds of particular individuals that continue in the narratives. And empires generally crumble along with their heroes.

In modern industrialized companies, we find embodiments of the red level hero characterized as "strong" business leaders, in whose companies, after their departure, missing structures and risks are revealed. See also [27] regarding this. This book shows how average companies differ from great ones, where calm and sober-minded CEOs do a clearly better job.

Level 4 blue: The level of existence, which is subordinate to absolute truth and where lasting peace is sought: absolutist, conformist, religious.

V-meme: Through a higher power, the world is given a structure and order that punishes "evil" and in the end rewards "good" deeds and proper moral conduct.

This level is characterized by **self-devotion** to a truth or a system, in order to receive rewards later. Those who belong to this system are committed to the bigger picture and willing to sacrifice themselves. Such individuals do not expect rapid fame, but expect to perish for a cause, either collectively or individually. Blue level individuals exist either as human manifestations or as perverted ideological content. Organizations with an active blue v-mem emphasize clear rules and reward belongingness and stability.

Level 5 orange: The modern, self-referring, "development-at-any-cost" level of existence: success-oriented, materialistic, objectivistic.

V-meme: The world is full of opportunities that can be used and develop to improve products and society to increase prosperity.

Those who win the game of the market are smart. This **self-expression** manifests without shame and without guilt when others are harmed.

This level, together with the blue level, dominates the modern industrial age. Since the Age of Enlightenment, the individual, rationality and development of progress and technology, along with all empirical sciences, have shaped social processes. Companies are predominantly competing according to "orange level" rules of the market economy, to be first, for the best stock exchange price, and for the highest profit. Many "business heroes" try to be winners in the field of big business. A loss of top position is equivalent to a loss of their entire existence. At the same time, this orientation has its dark side: Burnout on a personal level and economic crises on the collective level, are examples of limits of a purely rational, egocentric worldview.

Level 6 green: The pacifist, egalitarian "oriented–to-self-realization-of-the-whole-community" level of existence: personalist, group-oriented, humanist.

V-meme: The world is our home, which grows and thrives through mutual love, appreciation and participation.

Only fully-fledged members of the world, who are sensitive to themselves and to others, and who are part of the community. **Self-devotion** is human-oriented and rewards are attained by social recognition. The green level is based on several elements of the purple level. Last but not least, the concept of the modern heroic journey of Josef Campbell [26] comes from anthropology and research on the historical initiation rituals of clans.

In modern business culture, the sensitive hero only finds their place by resistance. Feminine leadership qualities, which are typical for the green meme, are required in our society, but due to the incorporation of other v-memes, at the same time they are destroyed.

Level 7 yellow: The existential, integrating level of existence.

V-meme: The world is a complex system in which change is constantly taking place and complexity is perceived as a gift: systemic, networked, integrating.

The "yellow" level individual has the meta-competency to be able to integrate into themselves seemingly opposing poles and levels and can transfer this to the outer world by adding value. Their **self-expression** is such, that attention is paid to all others in the system.

For individuals at this level of consciousness, Graves observed a sudden increase in problem solving competence, the integration of opposites and a high degree of autonomy, while at the same time respecting the limits of others. Graves therefore introduced a partition between the first six levels and the seventh level of his model. He named the first six levels, the "first tier" (the first section, the first order), and the seventh level and higher as the "second tier" (second order). Up until now, we have only had a rough understanding of how the yellow level is expressed in individuals and organizations. Opportunities for technical networking will create synergies and catalyze this v-meme.

We believe that the content of this book will help make the "yellow" level of existence more comprehensible and at home in project management.

Level 7 turquoise: The existential, integrating level of existence: experience-oriented, holistic thinking, synergistic.

V-meme: The world is an organism, which we have received as a gift and which transcends us. We bear responsibility, not only for humankind, but for the sustainable future of the entire universe. The self is ready to sacrifice itself, so that life can continue.

The yellow, and in particular, the turquoise "levels of existence" are at present, only slightly visible in parts. It is estimated that only around 0.1% of the world's population have this v-meme.

Individuals, groups, organizations, and societies, rather than possessing only one of the v-memes listed above, in most cases, contain a mixture of these v-memes. The degree of maturity, often referred to the "level of consciousness" of a person, group, organization or society, is measured by the balance of this v-meme mixture, as well as by the degree of penetration into all areas of life (contexts).

The 3 (red) to 6 (green) "levels of existence", correspond, in our opinion, directly to people's four basic needs as described in the appendices under "Fundamentals Consistency Theory" and "Fundamentals Reiss Motive Profile". We refer to those appendices.

7.6 Fundamentals Consistency Theory

In the section "Neuroleadership", we used Grawe's consistency theory to demonstrate leadership based on basic human needs. For the sake of completeness and to explain the Consistency Theory with respect to the Dilts Pyramid we will repeat parts of this description.

In his book "Neuropsychotherapy" [28] Grawe presents the latest findings from neuroscience, specifically for psychotherapists, and examines the question of which basic patterns of mental processes exist to help understand human action and to explain and solve mental disorders.

Based on a multitude of neuroscientific findings and the theory of basic needs, Grawe developed his "consistency theoretical model of mental functioning". According to this, people strive for compatibility (consistency) of different mental processes based on their basic needs. These mental processes are represented by neuronal processes. We strive to bring these basic needs in line by means of so-called motivational schemes, using our experience and behavior. Motivational schemes are specific behaviors acquired through experience. They can be differentiated by approach and avoidance schemes. As a result of ongoing experience, they can be reinforced or inhibited by other experiences. We combine motivational schemes with motives: For example, we bundle the approach schemes "I like to be the center of attention", "I want to set the tone" and "I push my interests through" with the motive "power". For this reason, we also accept simplification, and assign the motives proposed by Reiss to our basic needs (see also appendix "Fundamentals Reiss Motive Profile").

According to Grawe, there are four basic needs: the need for affection, the need for orientation and control, the need for self-appreciation and self-protection, and the need for pleasure and avoidance of pain.

- The need for affection is our need for human attachment and closeness. The motivational schemes associated with the need for affection are acquired, to a large extent, in the first years of our life. We assign the motives social contact and family to this basic need. In a professional context, the need for affection is also linked to ability within team orientation, as well as to the ability to establish confidence and consensus in professional collaborations and cooperations.
- The need for orientation and control is very strongly linked to our interaction with reality and our perception of reality. Our experiences teach us if and how we will achieve our goals. By making positive or negative control experiences, goal-oriented activities and control are inseparably linked to each other. The need for orientation and control is also linked to the desire for freedom of action and security. For example, we assign the motive "save" as a category of motivational schemes to the basic need for orientation and control. The motive "save" can combine very different motivational schemes. People for whom "save" is important, have at some time in the course of their life, learned that "save" helps them to control their resources and subsequently, there is room for other actions. Our ability to learn is also closely linked to the need for orientation and control. According to the central neural mechanism of learning, stress is important for learning, but only if we have it under control (see also section "Learning and Meta-Competency").

  We assign the motives independence, order, saving, idealism, honor, purpose and goal orientation, to the need for orientation and control. We would like to point out that by using the motive independence as an example, the bundling of motivational schemes to motives and values carries the risk of nominal distortion in itself. In the appendix "Fundamentals Reiss Motive Profile" we have mentioned that in the literature on the Reiss Motive Profile, independence has been replaced by team orientation. If e.g. "independence" is connected to the motivational scheme "It is important to me that I make my decisions alone", and "team orientation" is connected to "It is important to me that I know my decisions are backed by the team", then it is quite understandable for "independence" and "team orientation" to be viewed as two poles of one single motive. If "team orientation" is linked to the motivational scheme "I like to work in a team, because I like people around me", it makes more sense to assign "team orientation" to the basic need of affection. In this example, the limitations and dangers of the modeling of motivational scheme categories can be clearly seen.

- The need for self-appreciation and self-protection is considered to be one of the four basic needs, and a specific human need not found in animals. This is because this need requires an awareness of oneself as an individual and the ability for reflective thinking. All humans have the need for self-appreciation as an approach scheme, even when self-appreciation is abandoned for the purpose of satisfying the needs of others. For example, the need for self-appreciation is present even when satisfying the need for affection necessitates self-humiliation.
- Ultimately, we expect appreciation through affection. If important personal characteristics of our self-image are questioned by the social environment, statements detrimental to self-appreciation are "forgotten" using avoidance schemes for self-protection. Self-appreciation and self-protection are closely linked to cognitive distortions [28]: "It is mentally healthy individuals who have a distorted perception of reality with regard to themselves, and not those with poor mental health."[2] In addition, unrealistic optimism with respect to self has roots in the fact that [28]: "The probability of someone having a serious accident is considered to be much lower than it actually is."[3] We have assigned the Reiss motives power, status, recognition, emotional tranquility and vengeance/competition to this need.
- The need for pleasure and avoidance of pain, is an expression of our strive for pleasant states and our attempt to avoid unpleasant states. We constantly evaluate our experiences with respect to the qualities "good-bad". What is perceived as pleasant or unpleasant, is to no small extent, learned. Tastes and odors, among other things, are learned preferences: Bitter (including beer, coffee) and spicy (including chili) foods, for example, can become pleasant foods as a result of socialization, i.e. experienced with pleasure. We have assigned the Reiss motives curiosity, eating, romance and physical activity, to the need for pleasure and avoidance of pain.

In Fig. 7.7 we have assigned the most important terms of Grawe`s Consistency Theory to the neurological levels of the Dilts Pyramid.

In the above model, we have assigned the four basic needs to the neurological level "Identity". The motivational schemes have evolved through the interplay of biological preference and of approach and avoidance schemes acquired by experience. They protrude from the Identity level into the level of Values and Basic Assumptions.

Depending on an individual's perception and the reaction of the environment to behavior, intended objectives and results are evaluated either as coextensive (congruent) or not coextensive (incongruent). The more frequently (situation-dependent) behavior leads to (temporary) satisfaction of needs, the more likely it is that motivational behavior schemes will arise, i.e. an objective and its means of realization, from which a motivational scheme arises, is generalized.

[2] Translation by the authors.

[3] Translation by the authors.

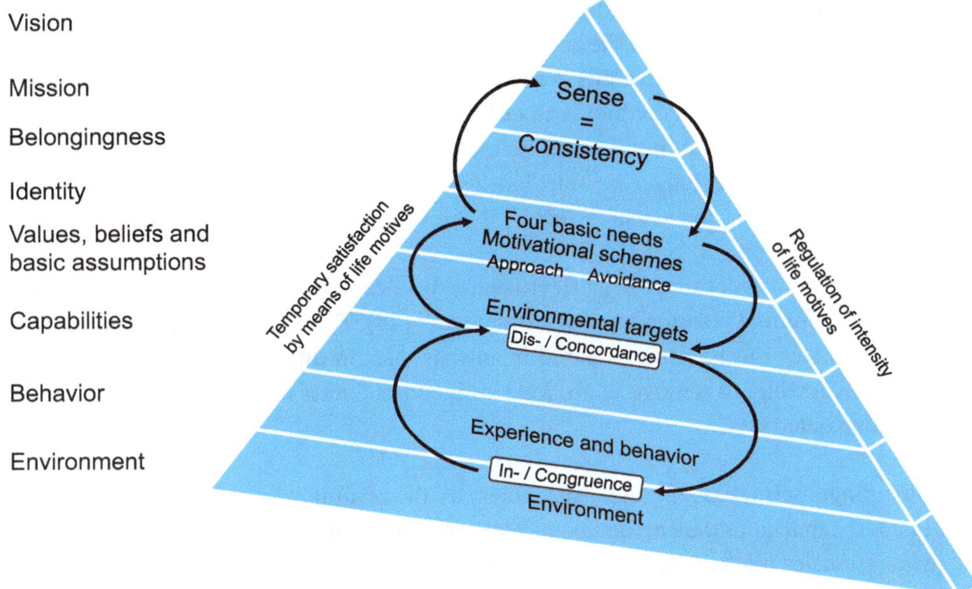

Fig. 7.7 Consistency theoretical model by Grawe, represented by means of the Dilts Pyramid

An approach scheme evolves, if a conserved and generalized experience is developed, which is obviously suitable for further successful need satisfactions. For example, an infant may have learned through experience that they can attract their mother's attention by shouting. The infant experiences their objective and actions as congruent. From this, a generalized experience is acquired, that the environmental objective, namely to get control over the mother, was successful. The environmental objective is in accord (concordance) with the basic need for control and order.

The counterpart to the approach scheme is the avoidance scheme. An avoidance scheme does not lead to the satisfaction of basic needs, but serves to protect them. A well-known example is when a child touches a hot griddle. Driven by the motive "to get to know the world", the griddle is explored. From the resulting experience, initially, only an avoidance scheme of pain avoidance is produced, namely not to touch the hot griddle in the future. If further such experiences are made on the basis of the motive "to get to know the world", this may significantly affect the pleasure of exploring, since further avoidance schemes are connected with this motive.

Research has shown that approach schemes and avoidance schemes are represented at the neural level by different neural circuits, which are independent from each other. We have already seen, elsewhere in this text, that past experiences cannot simply be "reversed". Schemes cannot be deleted, but only inhibited by other schemes.

One of the consequences of this is the potential to activate approach and avoidance schemes that are not compatible. In addition, simultaneously activated approach schemes may be incompatible. In this way, the "I want to assert my interests" approach scheme, can enter into competition with the "I do not want to be hurt by resistance in the team" avoidance scheme. We are all aware of the conscious and partially unconscious struggle between the different "selves". Because of this, Schulz von Thun [11] also speaks of the parts of our personality, which can be in conflict with each other. According to Grawe, these parts are nothing more than different motivational schemes.

If basic needs can no longer be satisfied due to competing motivational schemes, an inconsistency occurs at the "system level", as Grawe says. This is the basis for various mental disorders.

Inconsistencies undermine the effectiveness of dealing with the environment, i.e. endanger the "optimal positioning of the organism" [28]. Therefore, an individual's mental system evolves mechanisms to avoid inconsistencies in different circumstances or, if they are unavoidable, to remove them. Indeed, inconsistencies can result in two opposite and effective outcomes:

- In the longer term, existing inconsistencies, by the resulting prevention of satisfaction of needs, can lead to the emergence of patterns of mental illness.
- Inconsistencies that can be handled by the individual are part of a central neural learning mechanism.

A typical symptom of inconsistencies is a stress reactions. A mental stress situation occurs if individual perception of the achievement of objectives and the expected objective aimed for, are not coextensive (congruent). Basically, this congruence is also achievable by change of subjective perception. An example of incongruence is the purchase of a supposedly reliable car, which exhibits faults within the first few days after purchase. The buyer can take a general view of the purchase as a mistake (maximum incongruence), but can also view the faults as statistically possible, but unique events (reduction of incongruence). It should be noted that this pattern of the change of subjective perception is a proven behavioral therapy and is referred to in NLP as "reframing" [9].

Subjective evaluation of the manageability of the situation and duration of the stress reaction decides whether controllable stress ("challenging zone") or uncontrollable stress ("overdemanding zone") arises.

Controllable stress leads to specific excitation of the brain: The brain becomes more ready to learn, i.e. synapses can be established more easily, and an increased uptake of glucose takes place, resulting in increased metabolism. With quantitatively and qualitatively improved resources in this "state of alert", a "solution" is sought that reduces the level of incongruence. For this purpose, on a neuronal level, suitable neural circuits (excitation patterns) are formed by fluctuation between different possible states. We have already encountered this phenomenon several times. Possible excitation patterns are established with the resources, but immediately disintegrate again, if they do not lead to a reduction of incongruence. If a suitable neural pattern is found, it ends the stress reaction and the individual is back in normal mode ("comfort zone"). We have already seen this in the section "Regulation of Complexity Through Targeted Interconnectedness and

Self-Organization": The system looks for a new order parameter, if it is found, stability occurs because the "solution" is suddenly there. If the brain is unable to find one of these specific patterns over a longer period of time, an inconsistency is present in the system and mental disorders develop.

If a neural pattern repeatedly leads to a reduction of incongruence, it becomes part of a person's "solution repertoire". In future, in similar environmental situations, a stress reaction will not occur, as the "new solution option" (motivational scheme) prevents incongruence. The newly formed neural pattern has effectively become part of that person's skills, a learning process has been successfully completed.

Controllable incongruence, therefore limited in respect to intensity and duration, and caused by complex and diverse challenges, by exploiting fully its neural and genetic potential, leads to an equally diverse solution repertoire in a person.

Since people differ in respect to their genetic potential and experience gained in their lives, their ability to avoid inconsistencies in complex situations, and therefore stress reactions, is very different. Different personality types have different stress levels when entering a challenging and over-demanding zone. As a result, these two stress levels become further control parameters for individuals and subsequently, also determine the control parameters in the team. In the section "Regulation of Complexity Through Targeted Interconnectedness and Self-Organization", we became familiar with the "storming" team phase as the stress phase of the team. This phase must be carried out with controllable stress, otherwise the team will break apart at this stage. In the section "Transformation Management", we established a link to the Satir phase "chaos" of transformation work. Transformation work is stress and therefore only possible, if the "chaos" phase exists and is followed by the "integration of the new" phase.

An uncontrollable stress situation occurs if the subjective perception of objective achievement differs over a longer period of time to a large extent from activated objectives. This essentially results in two reactions by the brain. First, the neural excitation patterns already underlying acquired behavior (motivational schemes), which are obviously not appropriate in this case to reduce the level of incongruence, are replaced by seemingly more effective avoidance schemes. For this, a destabilization of the already existing neural connections takes place and thus a destabilization of neural and mental activity – exactly the opposite to the example of controllable stress. Second, the altered biochemical conditions caused by permanent stress may lead partially to considerable damage to certain areas of the brain. In summary, it can be stated that there is a high level of probability that uncontrollable stress leads to mental illness.

The activation of motivational goals is always accompanied by the activation of a reward system. Subjectively perceived goal achievement leads to positive feelings via biochemical processes, and non-achievement of goals leads to negative feelings. According to Reiss (see also section "Fundamentals Reiss Motive Profile"), positive feelings ("feelings of happiness") signal an occurrence of temporary need satisfaction by reaching a goal, and the related strengthening of a motive and value (value-based happiness). But in an ideal case scenario, evocation of these feelings is not the motivation. If the positive feeling itself is the motivation (feel-good happiness), the feeling is only of very short duration and intensity decreases each time the same objective is achieved again. A classic example is the "adrenaline junkie". In order to achieve the same intensity of feeling again (released adrenaline), they have to expose themselves to ever greater risks.

Avoidance schemes (e.g. "I do not want to fail or embarrass myself"), in contrast to approach schemes (e.g. "I want to finish the project successfully"), use more resources, because they tend to cause mental tensions by their negative objectives and are less focused. This reduces the resources available to approach objectives. Need satisfaction through value-based happiness is only attainable permanently by approach objectives. For this, it is necessary that the potential of an individual (their strengths) is consistent with the challenges of the environment at the relevant time.

Edelman and Tononi [29] formulated the thesis that a stable state of consciousness emerges in the brain due to an ongoing integration of information, included in distributed neuronal groups. This integration is an ongoing self-organized process. The sequence of these individual states of consciousness results in our perceived continuous consciousness. In order to be able to create each individual state of consciousness, a very large number of information states is integrated into a single coherent self-contained state of consciousness in a very short time. The degree of integration, i.e. the consistency, corresponds to the information content, the difference that makes the difference.

It is easier to understand this by looking at an "ambiguous image" that has two possible interpretations. In each state of consciousness, only one of the two "interpretations" is perceptible and the states of individual neuronal groups representing a possible second interpretation are hidden during integration in favor of an unambiguous perception. If we repeatedly focus on the image, we receive one of the two possible interpretations in our consciousness in a non-predictable way. The emerging macro-state, the state of consciousness, is located at the edge of stability [30], enabling the emergence of new states of consciousness, which are consistent with the previous states, without delay. Consistency is therefore an essential prerequisite for the functioning of the brain, both on the neural as well as on the psychological level.

7.7 Fundamentals Reiss Motive Profile

In his book "Who am I" [31], Reiss introduces for the first time, the concept of the sixteen motives of life ("Basic Desires"), which are also referred to as "Life Motives" and "Psychological Needs" in [32]. With [33], a German version of the Reiss profile is available from his near namesake Reyss and Birkhahn. Similarly to Grawe (see also appendix "Fundamentals Consistency Theory"), Reiss also considers the search for psychological needs that exist in all human beings, to be an empirically answerable question.

While as part of the research history of human needs, philosophical or psychological thought constructs were often used as a basis, Reiss conducted an empirical survey of more than 25,000 adolescents and adults in North America, Europe and Japan. The questionnaire used here is the Reiss Motivation Profile (RMP), which contains sixteen equivalent Life Motives (Table 7.3). The selection of the sixteen motives used in the RMP took place over several stages, from data collection (question on possible motives) and analysis for smaller groups (several hundred), to the validation of the identified motives with the application of the RMP for thousands of people. The description of this approach is part of the publication of results, which, however, does not contain the underlying analysis.

Depending on the source, the naming of the 16 Reiss Life Motives (we use the term "motives" in this book) vary. For this reason, we have also included the nomenclature according to [33] in Table 7.3.In parts, the other term emphasizes an opposing motive: For example, in [33] "team orientation" is used, whereas Reiss describes this motive as "autonomy". In Table 7.3, we have made assignments relating to Grawe's basic needs. The

Table 7.3 The 16 Reiss Life Motives grouped according to Grawe's basic needs

Basic need (Grawe) Life motive (Reiss)	Explanation life motive
Need for self-appreciation and self-protection	
Power	The need to affect others and to lead
Status	The need for social recognition, especially in an elitist sense
Acceptance	The need for inclusion to strengthen self-image
Tranquility (Emotional calm [33])	The need for emotional stability by avoiding fears and risks
Vengeance/Competition (Revenge/Fighting [33])	The need to fix somebody or to compare with someone
Need for orientation and control	
Honor (Goal- and purpose orientation [33])	The need to be principled and to be loyal to social environment
Independence (Team orientation [33])	The need for self-responsibility and autonomy
Order	The need that indicates the importance of structuring and flexibility for a person
Saving (Saving/Collecting [33])	The need that indicates the importance of possessing and collecting things
Idealism	The need for social justice and fairness
Need for affection	
Social Contact	The need for social contacts of both quality and quantity
Family	The need to have children
Need for pleasure and pain prevention	
Curiosity	The need for novelty and knowledge
Eating	The need for enjoyment of food consumption
Romance (Beauty [33])	The need for sex and beauty
Physical activity	The need for activity, sports and fitness

assignment of motives to basic needs were made according to plausibility and evidence. According to Grawe, motivational schemes are the basic building blocks of motivation, and motives are merely a useful form of grouping these building blocks.

As we have already pointed out in the appendix "Fundamentals Consistency Theory", we have no "motives in our heads". We show behaviors described by motivational schemes. Motives are therefore nothing more than meaningful groupings (categories) of motivational schemes. The RMP questionnaire contains, accordingly, representative motivational schemes phrased as questions. Assuming that an RMP statistical evaluation has been carried out carefully, the question of validity of motives is reduced to the question of whether the motivational schemes, which are expressed in the questions, are representative. Our experience so far leaves no reason for doubt.

Each of the sixteen motives outlined by Reiss has two significant characteristics (Table 7.4): The intrinsic goal (the actual goal, in contrast to goals, which are only "means to the end" for other goals) and saturation intensity. Saturation intensity is the individually required amount, frequency or intensity with which a goal must be achieved in order to trigger a positive feeling ("feelings of happiness") in an individual, i.e. to signal a person's temporary satisfaction of needs. In this sense, intrinsic goals are not a "means to an end" for other goals, but always a "means to an end" for need satisfaction (according to Grawe, intrinsic goals are referred to as (intrinsic) environmental goals).

Table 7.4 Life motives, goals, emotions, intrinsic values, according to [32]

Life motive	Intrinsic goal	Positve emotion	Negative emotion	Intrinsic value
Power	Influence	Self-efficacy	Humiliation	Leadership
Independence	Autonomy	Freedom	Dependency	Self-reliance
Curiosity	Ideas	Wonderment	Boredom	Insight
Acceptance	Avoid criticism	Self-confidence	Insecurity	Self-esteem
Order	Structure	Comfort	Uneasiness	Stability
Saving	Collection	Prudence	Wastefulness	Frugality
Honor	Character	Loyality	Guilt	Duty
Idealism	Justice	Compassion	Outrage	Fairness
Social Contact	Friendship	Fun	Loneliness	Belonging
Family	Parenthood	The feeling of being needed	Burden	Children
Status	Social rank	Superiority	Inferiority	Reputaion
Vengeance	Self-defense	Vindication	Anger	Winning
Romance	Sexuality	Ecstasy	Lust	Sensuality
Eating	Relish	Satiation	Hunger	Nutritional value
Physical Activity	Exercise	Vitality	Restlessness	Fitness
Tranquility	Caution	Relaxation	Anxiety	Safety

The sixteen motives, according to Reiss, are present in all human beings; but the intensity of individually required saturation (weak, average, strong) varies. Through the "Dictionary of normal personality traits" [32], Reiss assigns 500 characteristic traits to the 16 motives (for example, analytical, anxious, and arrogant). This assignment can help a person to deal with their own motives.

Since basic needs (according to current knowledge) are genetically anchored, the achievement of the motives' goals produces only a temporary feeling of satisfaction (the reward systems are activated), at regular intervals a new goal attainment is required to be reached. In other words, according to Reiss, the motives motivate for a lifetime – hence the notion of Life Motives. If we understand motives as groups of motivational schemes, the long-term orientation is certainly given, but motivational schemes can indeed change, which is even necessary during individual transformation work.

Reiss distinguishes between "well-being happiness" and "value-based happiness". Well-being happiness is the "savoring" of positive feelings generated by the body for signaling that a goal has been reached. Not only is this of a short duration, but the intensity of feelings is also reduced with each further achievement of the same goal. According to Reiss, everyone strives for long-term value-based happiness. Value-based happiness comes from goal achievement itself, and not from the positive emotions triggered by it. The achievement of the goal, strengthens the value connected to the specific motive (Table 7.4) in the environment of the individual, and results not only in a temporary satisfaction of needs, but also creates a positive expectation of meeting these needs in the future. In the appendix "Fundamentals Consistency Theory", we have already emphasized this positive relationship to manageable stress.

"To find the right measure, not too much, not too little," this is what people try to do with their actions. Reiss has adopted this insight, originally derived from Aristotle, and in the RMP asks the importance and/or the necessary degree (intensity), by which the goal of the motive is perceived as achieved (positive feeling).

He distinguishes between a weak (unimportant), average and strong (important) individually needed intensity. According to Reiss, only important and unimportant motives affect the characteristic traits of a person, not average ones. Amongst other things, this becomes obvious in his study on relationships, in which he contradicts the thesis that "opposites attract" and favors the thesis "people of the same kind stick together". Couples with many common motives, classified by them as important or unimportant, tend to stay together longer. In addition, the opposite is also true, couples separate more frequently when one partner regards many motives important that the other partner views as unimportant. The relative difference between people's required saturation intensity (different expression of motives) is a frequent cause of conflicts.

The context in which a person lives is probably responsible both for the degree of expression of a need (saturation intensity) and for the type of satisfaction, or motivational schemes. In [32], Reiss explains this using an example from the "social contact" motive: If the availability of social contacts in a person's environment is sufficient at the amount required for them, no special activities will be carried out for further satisfaction. If, on the other hand, the availability of social contacts is insufficient, the individual will activate approach motivational schemes to bring about a need satisfaction.

This can result in this person being characterized as "extraverted" in their social environment. Alternately, too much availability of social contacts can activate avoidance schemes, and the social environment might characterize the resulting behavior as "introverted". This showcases the fact that the intensity of personality characteristics displayed in different contexts is perceived differently by the individual themselves and their environment.

In a society, the dominant motivational schemes of all individuals are formed as values and beliefs, that is, as order parameters of society. These patterns of behavior are passed on from generation to generation as a way of satisfying needs and form central elements of a culture. Groups of these behavioral patterns are passed on, analogous to genes, as culture-defining thoughts (memes). In the Spiral Dynamics model, these behavioral groups are referred to as "value memes" (v-memes) (see appendix "Fundamentals Spiral Dynamics").

7.8 Fundamentals Diamond Model

In the section "En Route to Complexity", we became familiar with the task of a project as one of the complexity drivers. For this reason, it is important to bring transparency to the task of the project. This transparency should have a threefold effect in the team and in the circle of stakeholders:

- clarify what the task is,
- make visible what the stakeholders and, in particular, the project team understand by the task,
- develop a common stakeholder perspective for the task.

In order to comprehend this required transparency, we need a model, a uniform metric, against which to measure tasks. The search for a uniform metric that supports the above goals has a long tradition. For an overview of different approaches of metrics, also called project-type models, we refer to [34]. Similarly to personality models or cultural models, project-type models differ in their type and number of dimensions. This is partly due to the fact that in the models complexity drivers are mixed with complexity areas and an improper distinction is made between static and dynamic complexity. We refer to the section "En Route to Complexity".

Models for project-types can, on the one hand, be used for portfolio analysis in multi-project management, and on the other hand, for individual project analysis. For the sake of simplicity, models for portfolio analysis often only have two dimensions, which are represented in the well-known 2 * 2 Boston Consulting Matrix presentation. We will come back to this again later on.

We will now outline the Collective Mind Method model described in [7], followed by Shenhar and Dvir's Diamond Model [35]: We will see that both models are similar, but originate from entirely different perspectives. The project-type model of the Collective Mind Method has been developed from fundamental considerations of a project's organizational temperament [36]; and in the Diamond Model, statistical patterns were analyzed, over a period of 15 years, starting in 1990, for approximately 600 analyzed projects from different industry sectors. Since the two models are alike, and the Diamond Model with its "diamond representation" has a more compelling representation, in this book, we speak of the diamond model when referring to both project-type models.

Let us start with an explanation of the typology of the task, which we modify from [7]. In order to explicitly induce the temperament of the still to be formed organization "project", we allow ourselves to be guided by the following basic assumption: A temporary organization is formed by people. The characteristics of the task of a temporary organization are essentially determined by people's perception filters. Since we model the perception filter according to the MBTI, we describe the project-type according to MBTI dimensions. So we typify the project according to the four MBTI dimensions with the "dichotomy" pairs E-I, S-N, T-F and J-P:

Degree of mission (E-I dimension): This temperament dimension indicates to what extent the task or solution has to promote itself to the stakeholders to achieve sufficient acceptance. The necessary degree of mission of a solution is therefore very much dependent on the number of people affected by the project and the individual impact it has in the respective field of work. The size and heterogeneity of the stakeholder circle and the stakeholder-based learning of new structures, processes and technologies play a central role. Depending on whether a large or small group of people is to be reached through a project, the project must have either an extroverted or introverted temperament.

Degree of innovation (S-N dimension): The degree of innovation of a solution is very much linked to the concepts of intuition, vision or originality. We speak of an innovation when the solution affects the life of a considerable number of people in a sustained manner (private life or working life). Through innovative structures, processes and technologies, entire work areas can be revolutionized. If the degree of innovation is low, if known structures, processes and technologies are used, then it is just ongoing development. Analogous to MBTI typology, we speak here of the innovation – development "dichotomy" pair. This "dichotomy" pair is represented by the N-S pair of MBTI typology:

Innovative projects correspond to an intuitive organization and are therefore assigned to the NF- or NT- temperament. Ongoing developments or standardized tasks require a high degree of structuring and they transfer what was already known to other fields of application. This is ensured by the SF- or ST- temperament.

In order to better assess the degree of innovation within a project, it is useful to highlight the project with regard to the following aspects: Does the solution have a highly visionary character and therefore potential for innovation? Does the solution release surprising associations in the observer? Can we speak of originality, which makes the solution appear in a new light? Does the solution lead to new, domain-related knowledge associated with the solution? Or is new methodological knowledge required for developing the solution? Are the stakeholders, not represented in the project team, so they can acquaint themselves with the new solution by learning? Does the solution itself open up new possibilities of learning in the domain?

Degree of abstraction (T-F Dimension): We speak of abstraction when the task and the solution analyze and systematize by analytic-systemic intelligence (T-preference). In particular, complex task settings contain a large number of elements with a high degree of interconnectedness of elements. Once a solution has been developed intuitively, this solution has to be formed by abstraction.

In order to be able to better estimate the degree of abstraction within a project, it is useful to illuminate the project, for example, with regard to the following aspects: Does the task lead to complicacy or to complexity? Are there many components or aspects to integrate into a solution? Are there any interdependencies between subsystems? Is there a system of systems?

Degree of management (J-P dimension): The degree of management indicates the degree to which extent the project requires goal-oriented management. Goal orientation is required at all times, however, the stringency for tracking the goal (e.g. by deadline, quality and cost) depends very much in which phase the project is. A project in an initial phase needs more “room for maneuver” to find a solution than a project that is in the final stage and already oriented towards a defined goal. The project design has to adjust to this temperament, which corresponds to a J-P pair in MBTI typology.

Let us turn to Shenhar and Dvir’s model: In their book “Reinventing Project Management” [35], they propose a metric for the characterization of projects, which is structurally identical to that in [7].

They use the construction of the new International Airport of Denver, Colorado (USA) as an introductory example. As a result of an extended research visit in this region, one of the authors of the present book can attest to this project: The new fully automatic baggage handling system resulted in unintentionally opened or damaged suitcases. Hard-shell cases were recommended to be used whenever possible. How was this possible? An airport had been constructed and commissioned with a faulty baggage handling system? There is an explanation in Shenhar and Dvir’s book: While construction of an airport is similar to a complicated standard project, the new baggage handling system was a high-tech facility. Development and implementation corresponds more to an innovation project, yet it was implemented according to the methods of a complicated standard project, meaning that any emerging problems with the new technology could not be adequately resolved.

According to Wikipedia [37], this required a reworking of 16 months, so that the airport was only able to be operational after this period. The automatic system itself was only partly put into operation and finally switched off in 2015.

Shenhar and Dvir describe this by introducing the following dimensions for the characterization of projects:

Technology: This dimension describes how much new technology is needed to implement a project: Is established technology used? Are any new technological properties are used? Will completely new Physical Technology (PT) or Social Technology (ST) be used? Or will new PT or ST be applied in the project? This corresponds to the dimension, **degree of innovation.**

Novelty: This dimension describes how the project solution changes the market or the environment in which the solution is to be deployed: Is the solution based on known functions with minor changes; is a new combination of functions used; or is the solution completely new to the stakeholders. Stakeholders need to be won over for new solutions and to be carefully guided to them. This corresponds to the dimension, **degree of mission.**

Pace: This dimension characterizes how many constraints exist for the implementation of the project: Is there only a small amount of pressure, are there constraints in the environment, time, budget or resources, are the constraints critical; or is there a crisis with uncertain outcome? Management style needs to change, depending on these characteristics. This corresponds to the dimension, **degree of management.**

Complexity: This dimension describes whether the structures of the task favor complexity: Is there a manageable task with one or a few elements; has a system been created that contains several system elements? Or is there a system of systems with many interfaces? This increase in the complicacy of the structure of the task makes an increasingly systemic perspective necessary. This corresponds to the **degree of abstraction.**

Figure 7.8 shows the representation, chosen by Shenar and Dvir, with these four dimensions for Denver International Airport. The notations used in their book are the same as ours: This also means that we have applied the degree of innovation, not only to Physical Technologies (PT), but also to Social Technologies (ST) (in brackets, we have indicated Shenar and Dvir's axis names). The dotted, red line shows the characteristics of the subproject baggage handling and the solid blue line that of the airport itself. The name "diamond model" is derived from the similarity of a rhombus to a two-dimensional diamond.

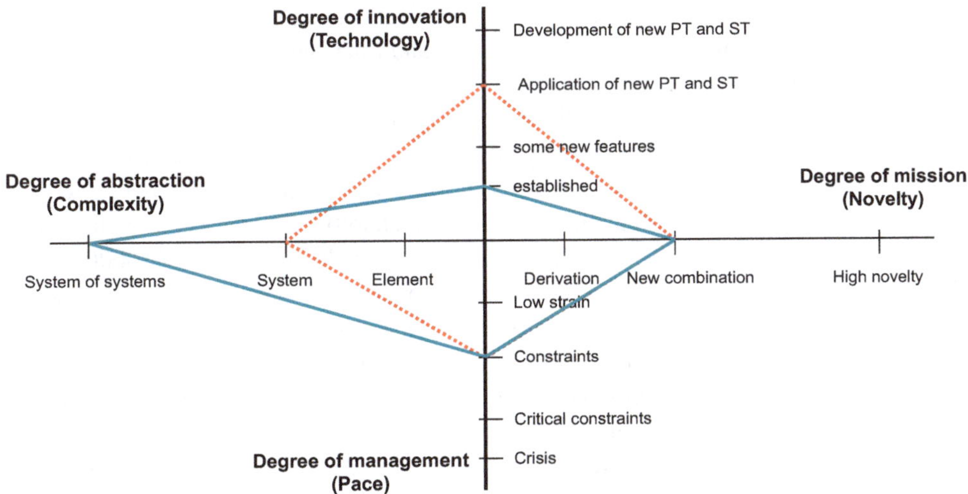

Fig. 7.8 The diamond model for the "Denver International Airport" project

The axes of the four dimensions contain Shenar and Dvir's classifications. In addition, it is useful to introduce a metric of 1 to 10 as in [7]. Such a metric facilitates the comparability of projects in a multi-project management organization.

In [38], the diamond model was discussed by proposing measures for the four dimensions and for each of the characteristics (i.e. the degree of innovation: some new properties, application of new PT and ST, development of new PT and ST). These measures are structured according to various perspectives. The proposed perspectives are integration management, content and scope management, scheduling, cost management, quality management, personnel management, communication management, risk management, and procurement management. For example, the question "What measures are necessary in the area of integration management, if the degree of innovation is at 10, i.e. the development of new PT or ST needs to be carried out"? As an alternative to these perspectives, we prefer the perspectives proposed in [39]: Social image, political image, intervention image, value creation image, development image, organizational image, change image. In our experience, these perspectives better support the focus and grouping of social success factors. If contemplation of the degree of innovation, results in consequences for the social perspective, these are directly integrated into the success factors of the "social image": For example, a high degree of innovation may require a special team setting and/or special team dynamics (see also section "The Big Picture" and chapter "Leadership in Complex Social Systems").

In the case of the "Denver International Airport" project example, the differences in the dimensions of technology and complexity between main project and subproject can clearly be seen. While the baggage handling system requires technologies that exceed those of the airport in their degree of innovation (dotted rhombus), for complexity, it is exactly the opposite case: An airport is more of a system of systems than the baggage handling system (solid diamond) is. For the stakeholders, both the main project and the subproject represented a new combination, i.e. novelty; for the project teams and later operators, this lay in the upper middle range. Since there were time and financial conditions, there was also a certain amount of pressure (constraints).

Shenhar and Dvir mention, in addition to many other examples, the development history of the Sony Walkman, which we would like to briefly present here. The Walkman came from the desire of the company owner of Sony, to be able to listen to music during air travel. His developers, with the help of known technology, designed the Walkman, as a favor for him, and it was produced as part of a first production batch of 30000 pieces. The Walkman was not accepted by the market, however, because people did not know what to do with it. This was the difficulty: In order to introduce the Walkman onto the market, it required innovative marketing concepts. While the dimension of innovation, the degree of abstraction (complexity) and degree of management (pace) were in the average range, the degree of mission (novelty) was different: It was much higher, as potential customers did not yet feel the need for this device. So, Sony employees all received a Walkman to use in public (for example in the subway), in order to make its attractiveness visible. The rest of the success story is history. The development and marketing of the Walkman can therefore be best described using a project in which the degree of innovation (technology) and the degree of mission (novelty) is high. We will call these projects below “Missionary projects”.

We recommend employing the diamond model at different times during the course of the project, in the stakeholder circle: This should be done during the initialization phase of a project, whether in the project team or as a means of communication with the customer, or in and with the sales team responsible for selling the project. During the course of the project, the model serves to make shifts visible in the scope. The degree of innovation may have changed, because new requirements have been introduced or because new technology is to be used with an invisible shift to move a project with a low degree of innovation to a project with a high or very high degree of innovation (technology). Even at the end of a project, the model can be used for a post-analysis of the project, for example, if, based on the course of the project, completely different perspectives are still expected amongst the stakeholders. In this case, such differences can be discussed in a Lessons Learned Workshop, or in a Retrospective, and serve to create a learning effect that goes beyond the project. In all cases, it is advisable that each team member firstly creates their own diamond model (to minimize any priming, and therefore, decorrelate), so that subsequently, individual assessments can be combined into an integrated diamond model. Insights, risks and measures are recorded directly, possibly according to different perspectives, on different colored post-its in the diamond model. Insights, risks and measures are included in the PDCA cycle for the assessment of success factors, as well as risk management and project planning and control.

In a similar way to the 16 MBTI personality types, which are simplifications, 16 types can also be derived from the diamond model.

From experience, we know that, in particular, degree of innovation (technology) and degree of mission (novelty) are underestimated, but often very important complexity drivers. For this reason, in a project portfolio, a first analysis according to these dimensions is particularly useful.

Figure 7.9 shows an example of a portfolio with five projects. In [7] we have defined four project types according to the four quadrants. To aid better memory recall, we have linked these four types of projects to the following professional groups: Inventor project, Missionary project, Builder project and Carpenter project. Following the text from [7] we describe these four project types as follows:

While the first two professional groups (inventors and missionaries) are visionary, sometimes volatile and eccentric, the last two professional groups (builder and carpenter) are characterized, above all, by accuracy and order. We can then transfer this metaphorically to the project types.

Projects that focus on innovation are called Inventor projects or Missionary projects.

The **Inventor project** (project P4) is a visionary, difficult, inventive task with a limited number of stakeholders. Interests are relatively homogeneous. It could be a task, like for example, the introduction of a completely new research method in the

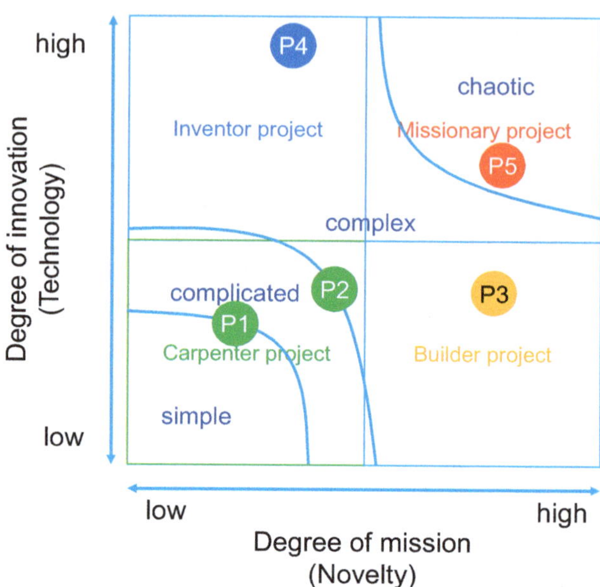

Fig. 7.9 Portfolio analysis

pharmaceutical industry. The task is also difficult, because many different aspects (for example, different process sequences) need integrating, and therefore a holistic approach is sought, for which new enabling technologies need to be applied. The solution is only used within a small circle. In a typical inventor project, the degree of innovation and abstraction is very high.

Missionary projects (project P5) is characterized by visionary, difficult, inventive tasks, with solutions being important for the larger stakeholder circle: If it is possible to implement a solution, for example, this changes the business model or changes the work environment of many employees so much that new ways of product invention are possible. In a typical Missionary project, a high degree of mission and innovation dominates.

Projects, where development in a known environment is the focus, we refer to these as Builder projects or Carpenter projects.

Builder projects (project P3) are structured tasks where solutions are important for the larger stakeholder circle: These are tasks, processed by using widely known methods and procedures, the solutions of which are hopefully conveyed to a larger community in a comprehensible manner. This includes, for example, the development of tailor-made user interfaces for SAP applications. In a typical Builder project, a high degree of mission dominates, closely followed by a high degree of abstraction and management.

Carpenter projects (project P1 and P2) are structured tasks for a limited number of stakeholders: These are tasks, processed using widely known methods and procedures, the solution of which are hopefully conveyed to a small stakeholder group in a comprehensible manner and implemented there. This includes, for example, setting up a Wiki for a small community. In a typical Carpenter project a high degree of management dominates.

The above figure is similar to the representation used by Boos and Heitger 1991 [34] where "social complexity" was plotted above "openness of the task", i.e. a complexity area against a complexity driver. If "social complexity" is associated with degree of mission, and "openness of task" with degree of innovation, the project types defined there, can be assigned to our project types: Carpenter project to standard project, Builder project to acceptance project, Inventor project to potential project and Missionary project to pioneer project.

In addition to project task typology based on the complexity driver's degree of innovation and degree of mission, Fig. 7.9 displays a classification according to the Wolfram classes of complexity (see also the appendix "Fundamentals Complexity Classes"). Ralph Stacey [40] initially introduced this representation in the context of the management of organizations. He used the axes agreement/disagreement (here degree of mission) and certainty/uncertainty (here degree of innovation).

The Stacey representation is very helpful in subsequent approximate assignment of projects, as long as the reader is not distracted into thinking that "project P3 is complex", but understands that "project P3 has a high level of probability of showing complex behavior." In the section "En Route to Complexity", we pointed out that in principle, every system is capable of showing simple, complicated, complex or chaotic behavior, according to the system parameters.

We have already pointed out above that the diamond model is appropriate for explaining the different stakeholder perspectives of the project. Discussion within stakeholder circles, using the diamond model, could be, for example, based on an expression of their different personalities, or, in other words, their different mindsets. The following examples illustrate this:

An example of different contexts (example taken from [7]): A company carries out a project and assigns the implementation of the solution to a contractor. The principal (the company) may conclude that the project has quite visionary and original temperament aspects: The principal's organization has never before carried out such a project and implementation of this project will have a considerable influence on its organization. On the other hand, the contractor may conclude that the project does not have any visionary aspects, because the resources and methods used are already familiar from other projects.

An example of different "belief systems": When discussing an IT project, the question arises as to whether the system to be created is a system with a high degree of abstraction (system of systems): All team members affirm that the system has approximately 30 interfaces to other IT systems. Some of the team members who would like to use new technology for the interfaces, argue that this could reduce the variety of interfaces to a maximum of five. This approach is too risky for other team members: They wish to use proven techniques, even if this means that 30 interfaces will need to be created and implemented. This example shows that different personal motives – willingness for innovation (adventure, curiosity, creativity) and safety – reveal different belief systems in the form of architectural principles.

7.9 Comment on NLP

NLP, or Neuro Linguistic Programming is a framework of methods, models, and theories introduced by Richard Bandler and John Grinder in 1975/76 by the two books "The Structure of Magic I/II"[4] [41, 42]. Both books are, like those of the greats of family therapy, system theory and hypnosis – Virginia Satir, Gregory Bateson and Milton H. Erickson – write, illuminators of their own professional patterns of thinking and communication.

NLP is now a framework of about 20 major methods and models that are applied in a vast variety of manifestations. NLP helps us to learn, become aware of, change and extend our own perceptions and expressions, as well as those of others, and automatic patterns of thinking and behavior. The basis of NLP is the modeling approach of Bandler and Grinder. The mental strategies ("neuro") of excellent personalities such as Satir, Perls, Bateson and Erickson or Einstein [43] are identified by analyzing their linguistic and nonverbal patterns ("linguistic"). The results are then transferred into comprehensible procedures ("programming"), so that these procedures can be communicated to others and applicable in other contexts [21].

NLP is now used in individual coaching, team coaching and organizational development (transformation management) [44, 8, 45, 9]. A simple search on the Internet shows that more than just a few psychological psychotherapists also use NLP as part of their therapy. – NLP is very similar to behavioral therapy.

In the 1970s and 1980s NLP was discredited by dishonest application in the field of sales. NLP was used for targeted manipulation in sales talks. Today these practices have disappeared to a large extent, especially as the corresponding code of practice by NLP (DVNLP in Germany) association has a regulating effect.

Evidence-based schools of thought assume that only statistically validated methods ensure (proven) effectiveness. These schools rely on specific, marginal models (such as the model of eye movement patterns) to discredit all other essential methods and models (such as the the Dilts Pyramid model or the meta-model of language). This is equivalent to the assertion: "Since the theory of ether is ridiculous the entire field of physics must also be discredited" (see also the appendix "Fundamentals Theory and Practice").

Despite the fact that experimental studies often do measure relationships, which are not asserted by theory, we do not believe in the exclusive criterion "evidence by statistics". Experience gained from numerous change projects has shown that NLP actually works.

Literature

1. Oswald A, Mourgue d'Algue H, Merkl G (2014) Fundamental requirements for a framework of adapted theories for project work. In: Rietiker S, Wagner R (eds) Theory meets practice in projects. GPM Deutsche Gesellschaft für Projektmanagement e.V., Nuremberg, pp 52–65

[4] Frank Pucelik is the third co-founder.

2. Popper K (2000) Vermutungen und Widerlegungen. Mohr Siebeck, Tübingen
3. Unzicker A (2015) Einsteins verlorener Schlüssel, CreateSpace Independent Publishing Platform
4. Santa Fe Institute (2014) Complexity explorer. Online Kurs Introduction to Complexity. http://www.complexityexplorer.org/courses. Accessed 20 Sept 2014
5. Stoica-Klüver C, Klüver J, Schmidt J (2009) Modellierung komplexer Prozesse durch naturanaloge Prozesse. Vieweg+Teubner Verlag, Kindle Version
6. Snowden DJ, Boone ME (2007) A leader's framework for decision making. Harv Bus Rev (November): 1-8
7. Köhler J, Oswald A (2009) Die Collective Mind Methode. Springer, Heidelberg
8. Dilts RB, DeLozier J, Bacon Dilts D (2010) NLP II – The next generation: enriching the study oft he structure of subjective experience. Meta Publications, Capitola
9. Mohl A (2010) Der große Zauberlehrling. Das NLP-Arbeitsbuch für Lernende und Anwender. Teil 1 und Teil 2. Junfermann Verlag
10. Stanford Encyclopedia of Philosophy (2016) Principia Mathematica. http://plato.stanford.edu/entries/principia-mathematica/. Accessed 26 Apr 2016
11. Schulz von Thun F (2008) Miteinander reden: 1, 2, 3: Allgemeine Psychologie der Kommunikation. Sonderausgabe der Ausgabe von 1981. Rowohlt Taschenbuch Verlag, Hamburg
12. Reynierse JH (2012) Toward an empirically sound and radically revised type theory. Journal of Psychological Type 1:1–25
13. Keirsey D (1984) Please understand me. Prometheus Nemesis Book Company, Del Mar
14. Keirsey D (1998) Please understand me II. Prometheus Nemesis Book Company, Del Mar
15. Keirsey.com (2016) Keirsey temperament sorter. MBTI Fragekatalog nach Keirsey. www.keirsey.com. Accessed 14 Feb 2016
16. Hossiep R, Mühlhaus O (2015) Personalauswahl und –entwicklung mit Persönlichkeitstests. Hogrefe Verlag, Kindle Version
17. Oswald A, Köhler J (2010) Wechselwirkende Organisationen, Teil 1. projektManagement aktuell 5:14–19
18. Briggs MI, Myers PB (1995) Gifts differing: understanding personality type. Davies-Black Publishing, Moutain View
19. Jung CG (2001) Typologie. Deutscher Taschenbuch Verlag
20. Charvet SR (2007) Wort sei Dank: Von der Anwendung und Wirkung effektiver Sprachmuster. Junfermann Verlag, Paderborn
21. Landsiedel S (2012) Trainingsunterlagen NLP-Master Ausbildung. Landsiedel NLP Training
22. Wikipedia (2015) Tangram Puzzle. https://de.wikipedia.org/wiki/Tangram. Accessed 14 Feb 2016
23. Keirsey D, Bates M (1990) Bitte verstehe mich. Prometheus Books
24. Beck DE, Cowan CC (2007) Spiral dynamics: leadership, Werte und Wandel. J. Kamphausen Verlag & Distribution GmbH, Bielefeld
25. Blackmore S (2005) Die Macht der Meme oder die Evolution von Kultur und Geist. Elsevier Spektrum Akademischer Verlag, Munich
26. Wagner U, Schwarzer S, Oswald A (2012) Der Helden-Begriff im Spiegel der menschlichen Bewußtseins-Entwicklung. In: Schildhauer T et al (eds) Realität und Magie vom Heldenprinzip heute. Verlagshaus Monsensstein und Vannerdat OHG, Münster, pp 98–105
27. Collins J (2001) From good to great: why some companies make the leap…and others don't. Harper Business
28. Grawe K (2004) Neuropsychotherapie. Hogrefe, Göttingen
29. Edelman GM, Tononi G (2013) Consciousness: how matter becomes imagination. Penguin Press Science, Kindle Version
30. Haken H, Schiepek G (2010) Synergetik in der Psychologie: Selbstorganisation verstehen und gestalten. Hogrefe-Verlag, Göttingen

31. Reiss S (2000) Who am I?: the 16 basic desires that motivate our actions and define our personality. Penguin Putnam, Kindle Version
32. Reiss S (2008) The normal personality: a new way of thinking about people. Cambridge University Press, Kindle Version
33. Reyss A, Birkhahn T (2009) Kraftquellen des Erfolgs. Mankau Verlag GmbH, Murnau
34. Dierig S, Witschi U, Wagner R (2007) Welches Projekt braucht welches Management? Sechs Dimensionen zur Projektdifferenzierung. Beitrag für das 24. Internationales Deutsche PM-Forum 2007
35. Shenhar AJ, Dvir D (2007) Reinventing Project Management: The Diamond Approach to Successful Growth & Innovation. Mcgraw-Hill Professional, Boston
36. Bridges W (1998) Der Charakter von Organisationen. Hogrefe-Verlag, Göttingen
37. Wikipedia (2015) Denver international airport. https://de.wikipedia.org/wiki/Denver_International_Airport. Accessed 15 Dec 2015
38. Haberstroh M (2013) Angepasstes Projektmanagement bei Unsicherheit und Dynamik. In: Wald A, Mayer T-L, Wagner R, Schneider C (eds) Komplexität. Dynamik. Unsicherheit. Advanced Project Management, vol 3. GPM Buchreihe Forschung, Buch Nr. F07. GPM Deutsche Gesellschaft für Projektmanagement e.V., Nuremberg, pp 174–195
39. Winter M, Szczepanek T (2009) Images of projects. Gower Publishing
40. Stacey RD (2015) Strategic management and organisational dynamics: the challenge of complexity, 7th edn. Pearson Education, Kindle Version
41. Bandler R, Grinder J (1975) The structure of magic I, a book about language and therapy. Science and Behavior Books, Palo Alto
42. Bandler R, Grinder J (1976) The structure of magic II, a book about language and therapy. Science and Behavior Books, Palo Alto
43. Dilts RB (1992) Einstein: Geniale Denkstrukturen und Neurolinguistisches Programmieren. Junfermann Verlag, Paderborn
44. Brinkmann M (ed) Besser mit Business NLP. DVNLP e.V., Berlin
45. Giesen R, Kluczny JW (eds) (2011) Coachingperspektiven. Impulse für die Praxis. DVNLP e. V., Berlin

Glossary

Abstraction Abstraction is understood (here) to be a form of bundling or category creation: For example, different behaviors are assigned to an ability, capabilities are in turn associated with a value or a basic assumption, etc ... Or, a system element is assigned to a system; a system is associated with a system of systems. Here, with each level of bundling or category creation, the bundles or categories become more abstract, since without further knowledge they can no longer be inferred by their specific context.

Actuality Actuality is the part of reality which has an effect on us and which we can observe.

Agile Organization An Agile Organization is present if the mindset of the organization allows a high level of flexibility and speed in adapting to its environment.

Antifragility Antifragility refers to a system property, in which the system is not broken by shock, but becomes stronger. To make this possible, the impact of the shock to the system may only be large enough for the system to adapt to learning. For example, a business is fragile, if the enterprise is primarily based on standardized processes and has no, or only a few mechanisms of adaptation, so that when large changes in customer behavior occur, there is little chance of survival. Fluid organizations are antifragile.

Attractor An attractor is a state of a system in a phase space defined by one or more order parameters.

Axiom An axiom is a statement that serves as the basis of a model or theory and is not derived or proved.

Basic assumption Basic assumptions are principles, underlying assumptions or belief systems that as a result of processed experiences, represent generalized mental models for the regulation of complexity. Basic assumptions act as situation-specific order parameters.

Bias A bias is a mental distortion, which leads perception in a specific direction.

Change Management By Change Management, we understand organizational change work, in which the starting point, the destination, and the path from source to destination are known. The change agent acts primarily as a manager who deliberately leads the organizational system in a specific predetermined direction.

A. Oswald et al., *Project Management at the Edge of Chaos*,
https://doi.org/10.1007/978-3-662-48261-2

Chaos A system displays chaos if it forms endless irregular states. The adopting of these states is assigned deterministically and depends heavily on the initial conditions of the system.

Coherent state A coherent state refers to a system state, in which the system elements form an internally consistent, interrelated and comprehensible whole.

Cohesive state A cohesive state refers to a coherent state, which is maintained by the system elements over time. The Collective Mind state is a cohesive state.

Coincidence Coincidence or coincident event is an event, the cause of which is unknown or not revealed to us.

Complex equivalence A complex equivalence exists if the behavior of a communication partner is unconsciously interpreted in the light of our own injured value or motive.

Complexity By complexity we understand the mechanisms of individual elements of the micro-tier of a system, based on a diversity of behavioral alternatives and entails a variety of behavior options at the macro-tier. We perceive complexity if

- a high degree of interconnections in time and/or place is present in a system,
- small changes have big effects,
- and erratic, inexplicable system behavior in time and/or place may occur.

Complexity domains Complexity drivers induce complexity in specfic domains of complexity. We distinguish between the two major complexity domains: "social complexity" and "solution complexity".

Complexity drivers Complexity drivers are factors that induce complexity. In projects, this includes: the scope, stakeholders and social organizations, as well as the wider environment.

Control parameter A control parameter is a parameter that allows us "to conduct" a system in different areas of the formation of emergent properties.

Correlation Correlation describes the interrelation between two or more features, events, states or functions. It does not describe a cause-effect relationship, but is a measure of the degree of pattern and structure formation in space and time. For example, a correlation function may indicate the likelihood that in a meeting, two people display the same behavior within a specific time interval.

Culture By culture, we understand the predominant mindset of an organization and model according to the Dilts Pyramid.

Decorrelating measures Decorrelating measures serve to prevent the interrelationship between two or more features, events, states or functions. For example, it may be necessary to prevent, or at least minimize, the interrelation of behavior of two or more people. This is especially true if mental biases occur in a team.

Dynamic Each model of a system contains, in addition to statics, statements on the dynamics, i.e. how the variables, describing the model, develop over time and space.

Emergent state An emergent (macro-) state is created by the mechanisms of complexity, based on the interaction of individual system elements. Strong emergence is caused by self-organization in a complex system, creating a new cohesive state on macro-tier, caused by "enslavement" of the individual system elements of the micro-tier.

Emergent properties Emergent properties are macro-tier properties of a dynamic system, which are neither present in the individual system elements of the micro-tier, nor inferable therefrom. Water molecules are not moist. Water itself has the weakly emergent property to be moist. Neurons cannot think. Our brain has the highly emergent property to generate consciousness.

Emergence Emergence is understood as the creation of emergent properties at the macro-tier of a system.

Empathy Empathy refers to the ability and willingness to identify, understand and empathize thoughts, emotions, motives and personality traits of another person.

Empiricism Empiricism relates to the specific observation of actuality, of gained information and methods for obtaining this information.

Entropy Entropy is a measure of the number of micro-states, representing a macro-state. Thus, entropy is a measure of "missing information": It is a measure of the information that is missing in a macro-state compared to a micro-state. For a solid body, the entropy is smaller than for a liquid body, because for a solid body, the number of micro-states which represent the "solid" macro-state are fewer than the number of micro-states which represent the "liquid" macro-state. Due to the interaction in the solid body, the degree of freedom is reduced, whereas the number of possible micro-states decreases. In a complex system, which is subject to self-organization, therefore forming a dynamic macro-state, entropy is also smaller than in a system without self-organization.

Evidence Evidence indicates a clear, coherent and obvious insight making a claim to truth. In the adjective "evidence-based", this understanding is reduced to "instance-based".

Flow Flow refers to a psychological or social status, where challenges and skills undergo a continuous re-balance. The Collective Mind requires a flow within the team as a supreme control parameter.

Fluid Organization A Fluid Organization exists if a network of Agile Organizations, consisting of a permanent organization, task-related temporary organization and temporary project organization, dynamically adapts to its environment.

Hypotheses Models and theories model reality. Hypotheses are statements derived from models and theories relating to actuality and practice. The validity of the hypotheses must be checked against actuality and practice. Hypotheses that cannot be confirmed by actuality or practice, result in falsification of models and theories.

Insight Theories provide an understanding of the perceived (observed) actuality. They give meaning to actuality and therefore give order, which results in insight. They thus contribute to knowledge of the actuality.

Integral theory An integral theory consists of several theories and models, but its meaning is revealed only by contemplation as a whole.

Intervention An intervention refers to an input (e.g. stimulus, irritation, selection, change) in a system, with the aim of achieving a targeted reaction. Since the reaction of a complex system cannot be predicted, and because the input is influenced by an interaction of the input and the system, interventions are not subject to linear cause-effect mechanisms.

Leadership parameters Leadership parameters of a social system are the setting parameters, the control parameters and the order parameters of the system.

Learning phase Learning phase refers to a personal or organizational state, of being aware of one's own actual competences within an area of competences. There are four phases of learning: unconscious incompetence, conscious incompetence, conscious competence, unconscious competence.

Learning stages Learning stages refers to degrees of meta-competency. These are categorized into learning stages I-IV.

Macro-tier The macro-tier of a system describes the system level of a system; a macro-tier "abstracted" from a micro-tier. The system "water" is described at a macro-tier by system properties such as wet or liquid.

Macro-structure The macro-structure of a system comprises all emergent system properties. For the system "water", the system property "wet" belongs to the macro-structure. For the system "organization", culture belongs to the macro-structure.

Management Management is a social technology, which develops models and theories for designing and leading social organizations, and applies these as social techniques.

Meme Memes are specific pieces of information that act as replicators of information by means of variation, selection and heredity, and are drivers of information evolution. Values or beliefs are memes.

Meta-competency Meta-competency means perceiving and categorizing the characteristics of one's own personality, that of communication partners and that of organizations, masterfully along the respective mindsets. This means that not only is behavior observed but also the context in which this behavior is shown, and the related manifestations of higher levels of logic. It also means, if necessary, removing yourself (mentally) from the respective system and observing the system from an external perception position (meta-perception position).

Meta-program Meta-programs are person-specific perception filters. They are structures and patterns of our personality that determine our thoughts and actions. They determine to a great extent, how we distort, obliterate and generalize information. The most important meta-programs are facets in MBTI dimensions.

Method A method is a planned process that when applied in a certain context, results in the solution of theoretical and practical tasks. Methods are derived from one or several models and related theories.

Micro-tier The micro-tier of a system describes elements and their interactions. The system "water" (in a pail) consists at the micro-tier, of water molecules and their interactions.

Mindset A mindset is a set of characteristics that a person or an organization has at all logical levels of the Dilts Pyramid. For individuals, it is thus a model of inner attitude, the personality. For organizations, the prevailing mindset is culture.

Model A model is a representation of actuality. Models are often components of a theory.

Motivational scheme A motivational scheme is behavior that satisfies one of our basic needs.

Motive A motive is the name for a collection of motivational schemes, which have a high level of similarity.

Neuroleadership Neuroleadership is the design of an individual working environment, according to an employee's four basic needs. A manager, will therefore need to apply different organizational and personnel management measures.

Neurological level Neurological level refers to a level in the Dilts Pyramid. The levels are, from bottom to top: context, behavior, capabilities, values and basic assumptions, identity, belongingness, mission and vision. Logical level is a synonym for this level.

Observation Observations are perceived events and phenomena of actuality. Theories explain observations that are made in actuality. Observations relate to specific segments of actuality, these segments are consciously designed in experiments (trials).

Order parameter An order parameter is a system parameter, which induces and characterizes an emergent macro-state.

Organization An organization is a social system consisting of a group of people who share a common purpose and use common resources.

PDCA PDCA refers to a cycle of Plan-Do-Check-Act:

"Plan" is the formation of a success factors – success criteria network, followed by the selection and assignment of theories and models for the design of success factors; as well as the formulation of hypotheses on the application and effect of specific aspects of the theories and models in the network of success factors – success criteria. Example: Based on a success factor team, a model is selected for the team setting and a concrete team setting is modeled (a hypothesis is formed). Here, e.g. project type is included.

"Do" means applying the hypothesis. Example: The team is composed in accordance with the team model and in practice, monitored on effectiveness in achieving success criteria.

"Check" means checking network success factors – success criteria. To achieve this, success factors and success criteria are checked for meaningfulness. Furthermore, the theories and models and the derived hypotheses are highly scrutinized.

"Act" making corrections in practice, based on "check". Subsequently, the PDCA cycle leads in a continuous sequence of cycles until the end of the project.

Phenomenon A phenomenon is an event which is observed and associated with actuality.

Phase space The phase space desribes the set of all possible states of a system. Every state of the system corresponds to one point in the phase space. If a temporal dimension is added to the phase space, we speak of a state space.

Practice Practice refers to the context "theory and practice". Practice includes all actions, which are based on mental conceptions. These mental conceptions are based on implicit or explicit theories.

Principle Principles are explicit basic assumptions about actuality and are the basis of models and theories.

Projections Projections are predictions, which are based on models and theories. Projections can include recommendations to act in a certain manner.

Reality Reality describes the world "in itself". We perceive this reality as an impression, as actuality.

Regulation of complexity By regulation of complexity, we mean any modeling of the complexity of a system to influence the complex behavior of the system. For this purpose, well-considered stimuli or interventions to the complex system are exercised. For the social system "project", a system development aims to be in line with the project goal and objectives. There are four categories of regulations: Regulation by shielding in space and time, regulation by forming models and intuition, regulation via targeted mental networking and by social self-organization and regulation by organizational setting parameters, control parameters and order parameters.

Resonant communication Resonant communication exists if dialog partners are able to align their own behavior to that of the perceived mindset of the other partner.

Reward system Reward systems are complex neural processes with strong feelings (of happiness), accompanied by neurotransmitters and neuromodulators like serotonin and dopamine, amongst other things.

Self-organization Self-organization creates, under certain conditions (setting, control and order parameters) and through the interaction of system elements on the micro-level, emergent properties on a macro-tier. The consequence of a self-organizing process is an ordered pattern, collective behavior or a new structure.

Self-referentiality Self-referentiality refers to the ability of a system to create a reference for itself, in order to distinguish it from its environment. The ability for self-referentiality is an important precondition for self-organization.

Sense Creating sense, refers to establishing meaningful connections to one's past, to one's present and to one's future.

Social Technologies Social Technologies (STs) are comprised of a set of social techniques.

Social techniques Social techniques are practical tools, based on theories and models, with the purpose of organizing people towards a goal or a vision.

Stakeholder Stakeholder refers to all persons, parties or groups interested in a particular project. They are the ones who "are at stake".

Stakeholder Management Stakeholder management identifies, supervises and proactively organizes, the chances and risks related to the stakeholders, and their interactions, for the sake of the project.

State State denotes the entirety of all observed properties of an object at a specified point in time and space. A state is therefore not static, but characterized by dynamics.

Success factors Success factors are factors that influence success criteria. The design of success factors, such as team composition or leadership, is made by theories and models, and is continually monitored in a PDCA cycle.

Success criteria Success criteria are the criteria by which a stakeholder measures the success of a project. Success criteria cannot be directly influenced.

Synergetics Synergetics is the theory of self-organization, which is applied in physics, biology, chemistry, psychology and sociology. The theory relies on the so called "formalism of master-equation".

System A system is an aggregate of elements, which relate in such a way to each other, that they can be thought of as purpose-, goal- or task-oriented unity, and in that regard, distinguishes itself from its environment. Models and methods are purpose-, goal- or task-oriented perspectives on the practice of a sytem.

Systemic project management Systemic project management refers to a project management style, which focuses on a systemic perspective. This dominant role of a systemic perspective, is supported by the fact that complex systems can be characterized by a micro-tier and a macro-tier. The micro-tier is the level of single elements or agents in the system. The macro-tier is the level of emergent structures. Systemic project management therefore, sheds light on the following questions: Which interactions exist between micro- and macro-tier? Which manifestations of the micro-tier lead to which manifestations of the macro-tier? Which manifestations of the macro-tier lead to which manifestations of the micro-tier? Which techniques of intervention exist on both micro- and macro-tiers, in order to influence the system in a goal-oriented way?

Target hierarchy Target hierarchy refers to the 3-tier model of the Collective Mind scheme: Goal-tier, What-tier and How-tier.

Theory A theory designs an image of reality or actuality.

A theory consists of one or more models, and attempts to explain observations made in actuality. Here the theory is supposed to satisfy certain quality criteria: falsifiability, auditability, consistency, fertility, operationalizability. The theory not only enables an explanation of the narrow span of observation, but also creates projections about actuality. It enriches human insight, which contributes to an extension of our knowledge.

Thesis A thesis is an assertion which requires proof.

Trajectory A trajectory relates to the sequence of all states, which describe the development of systems, with reference to an initial point in time and space.

Transformation management Transformation management is an organizational change work, when most of the starting point is known, however, the goal and the path from source to destination is unknown and open. The transformation agent acts primarily as a coach who accompanies organizational transformation through systemic interventions

Transition management Transition management is organizational change work where the starting point and the goal are known or defined, but the route from the starting point to the goal is unknown and open. The transition agent acts primarily as an advisor who specifies guidelines for the system, but in accordance with those, leaves a lot of freedom.

Typology A typology is a "typecast" for a certain field of application and often involves simplification. Fields of application are, for example, personality type and projects. Initially, with personality type models, the term typology was understood as an expression of polarized personality preferences: for example a person would thereafter be an extrovert or introvert (e.g. two poles). Today, we know that our personality "types" are based on continua. For example, every person shares both extrovert and introvert attributes. We apply the term typology in accordance with this insight and therefore deviate from the historically shaped understanding, even when we frequently use the term "type" in explanations and descriptions for the sake of simplicity.

Variety Variety describes the quantity of states of a system, either on the micro-tier or the macro-tier.

Values Values, value ideas or ideals are based on processed experience and are generalized, emotional guidelines for regulating complexity. Values are personal, organizational or social benchmarks for social action and are the foundation of the cohesion and further development of a society. Values can function as order- and control parameters. Examples: Love, Friendship, justice, duty, discipline, loyalty, integrity, human dignitity, solidarity, liberty, tolerance...

Value-creating complexity Value-creating complexity exists if "the whole is more than the sum of its parts". A value-creating social complexity exists, if connections within the team or the organization result in communication processes which make "the whole more than the sum of its parts".

Value-destroying complexity Value-creating complexity exists if "the whole is less than the sum of its parts". A value-destroying social complexity exists, if connections within the team or the organization cause communication processes which make "the whole less than the sum of its parts".

Literature

1. Shakespeare W (2004) Hamlet: Bilingual edition (trans: Günther F). Deutscher Taschenbuch Verlag, Munich
2. Shakespeare W (2009) King Richard III: Bilingual edition (trans: Günther F). Deutscher Taschenbuch Verlag, Munich
3. Shakespeare W (2012) The Tempest: Bilingual edition (trans: Günther F). Deutscher Taschenbuch Verlag, Munich
4. Shakespeare W (2012) King Lear: Bilingual edition (trans: Günther F). Deutscher Taschenbuch Verlag, Munich
5. Shakespeare W (2014) A Midsummer Night's Dream: Bilingual edition (trans: Günther F). Deutscher Taschenbuch Verlag, Munich
6. Vargas F (2010) The Chalk Circle Man (trans: Siân Reynolds). Harvill Secker, London, Kindle Version
7. Vargas F (2013) The Ghost Riders of Ordebec (trans: Siân Reynolds). Penguin Books, New York, Kindle Version
8. Vargas F (2015) A climate of fear (trans: Siân Reynolds). Harvill Secker, London, Kindle Version

A. Oswald et al., *Project Management at the Edge of Chaos*,
https://doi.org/10.1007/978-3-662-48261-2

Zeitfracht Medien GmbH
Ferdinand-Jühlke-Straße 7
99095 Erfurt, Deutschland
produktsicherheit@kolibri360.de